A Journey to Heal

Also by Edward E. "Ed" Saunders

Sentinels, The History of Yellowstone National Cemetery from Prairie to Hallowed Ground (2016)

Knapsacks and Roses, Montana's Women Veterans of WWI (2018)

Forgiving Dad, An Autobiography (2018)

For the theater:
The Rule (A stage play in two acts) (2022)

A Journey to Heal

Photography and Reflection on the Nez Perce Trail

EDWARD E. SAUNDERS

Copyright and Library of Congress Information

A Journey to Heal, Photography and Reflection on the Nez Perce Trail

Published by Edward E. Saunders; Billings, Yellowstone County, Montana

Paperbound ISBN: 978-0-9973265-3-6
Interior and cover design by the publisher. Cover photographs: Lolo Trail, and Nez Perce rider at Canyon Creek battle site, Montana.

First edition
First paperbound printing, 2025
Printed and bound in the USA

1. Nez Perce 2. Nez Perce War 3. Chief Joseph 4. Montana 5. Montana History 6. Plains Indian Wars 7. Idaho 8. Idaho History

Type setting in Minion Pro, 11 pitch

To the warriors, Native and soldier,
of the American West long ago.

CONTENTS

ILLUSTRATIONS AND GRAPHICS

This book is primarily a visual book with many included photographs. Listing them individually here would create voluminous and cluttered pages of little value to the reader. For further explanation, each photograph has a summary explanation supporting the text of each respective chapter.

GRAPHICS

ACKNOWLEDGEMENTS

Dedicated people in the American West and southern Saskatchewan, Canada, with passion and foresight, preserve the rich history, heritage, and culture of the vast plains, hills, and mountains of the western continental ridge of the region.

I owe much to people who gave of their time to help and for their valuable review, critique, and constructive criticism of many aspects of this book: Alan S. Miller and Tracy L. Livingston, Corrales, New Mexico; Gary and Paula Durell, Billings, Montana; and Dr. Robert B. Pickering, PhD, Dean of Library, University of Tulsa; Tulsa, Oklahoma.

To Dave Wanzenried, Billings, Montana, whose passion and interest for historic preservation at the site of the Battle of Canyon Creek, Nez Perce War, kept me energized to complete this book. To Mike Penfold, Billings, Montana, who shares a remarkable passion for historic preservation in Montana.

A thanks to the staff and executive leadership of Hathi Trust Digital Library and to Google. Their foresight in putting online digitized historical reports, free-of-charge electronic books, and other voluminous print materials made my research easier and rewarding.

To the US National Archives and Records Administration for putting many digital photographs online for public use. To the Library of Congress, and the National Endowment for the Humanities for the *Chronicling America* digitized newspapers from America's history.

To the Montana State Library's *Montana Memory Project*, Helena, Montana, which has publically available print material, Montana newspapers, photographs, and other valuable contributions.

To the staff at Western Heritage Center; Billings, Montana, who share an undaunted passion for history.

To Region 1 of the US Forest Service, Hamilton, Montana, and Ms. Blakely D. White, who assisted in getting me permission to fly my FAA-registered and licensed small unmanned aircraft system in specific US Forest areas along the Nez Perce Trail.

To Mark Davison, National Park Service, superintendent, Fort Laramie National Historic Site; Fort Laramie, Wyoming, who gave guidance and clarification on flying small unmanned aircraft in National Park Service managed areas.

To the staff of the Nez Perce National Historic Trail for their help and assistance.

To the staff and director of the Fort Walsh National Historic Site; Fort Walsh, Saskatchewan, Canada. They took time to personally and passionately explain the engaging history of that important fort along the North West Mounted Police Trail.

To the Saskatchewan History and Folklore Society, Saskatoon, Saskatchewan, for their help in research. To the staff at Grasslands National Park, Parks Canada, Val Marie, Saskatchewan, for their help in directing me along the North West Mounted Police Trail and to Pinto Horse Butte, the site of Sitting Bull's camp.

To the manager and technicians of Standard Buick, Chevrolet, and GMC; Shaunavon, Saskatchewan, who stayed past quitting time to repair a blown tire on my vehicle to keep me going through the vast Saskatchewan prairies.

To the staff and director of the Baxter Springs, Kansas, Heritage Center and Museum. This is a remarkable repository of Nez Perce history when Chief Joseph and his followers were exiled to the Indian Nation.

To the staff of the Quapaw Nation; Quapaw, Oklahoma, who gave of their time to show me their historic collection. To the staff at the Tonkawa Nation administrative center; Tonkawa, Oklahoma, who ensured I knew of the Nez Perce Cemetery.

Lastly to my wife, family, and friends, who realized my commitment to this book and to preserve in contemporary photographs the legacy of a remarkable people caught in a tornadic swirl of conflict they did not create.

INTRODUCTION

To Heal A Wound

Gray clouds formed in early October 2022 as I walked barren and lifeless ground forty-two miles south of the Canadian border in north-central Montana. Cold rain came during my third trip to this place. The Canadian province of Saskatchewan, two easy days horseback-ride north, sent wind, always the wind, mocking a person's resolve.

In October the north Montana prairie turns light brown. For weeks no rain nourished the withering prairie grass. The grass had surrendered for now, but it would come again lush green in spring. Relentless summer sun had baked the ground, baked it hard. Streams, once full and clear from snowmelt months ago, now barely wept with what remained of life-giving water.

Winter comes early this far north, winter keeps its own calendar here. October should bring colorful pleasant autumn. Not here. Not this far north. Winter and its companion the cold, turn rain into sleet. The wind then hurls sleet like knives lashing skin and blinding the eyes.

Nothing would have sheltered anyone that day from winter's early assault. Leaning into the wind, head down, eyes strained, I labored the one-mile circular trail surrounding an austere 100-acre patch of prairie emptiness. No trees, shrub growth, or sheltering rocks filled that patch. Bear Paw Mountains on the southwest horizon did not block the weather: Weather that was turning gray, ugly, and angry.

The National Park Service now cares for and manages this troubled ground, deemed sacred and rightfully so. The Park Service carefully and with precision mowed the knee-high prairie grass forming the circular trail: a foot path above the prairie sod. The Park Service had placed colorful waist-high information signs along the trail explaining the emptiness and events, which transpired in history before you.

Your legs feel the elevation change as the trail goes uphill east of the emptiness. A lone bench greets you on the east high ground. If you

October 2022, empty and barren Bear Paw Mountain battlefield, Nez Perce War. View north. Snake Creek in middle. Nez Perce camp was in upper center. National Park Service commemorative path shown at bottom and curving around right center.

choose to sit on the bench, you face west overlooking a sluggish snake-like creek bed with rolling hills and dry rivulets feeding the creek. The creek has a disturbing and sinister name of Snake Creek.

In early October 1877, terrified screams, cries of anguish, and death wails caused from deadly rifle and cannon fire, filled with blood the same emptiness below you along the creek. Wind and rain came then too. The remnants of valiant nontreaty Nez Perce and equally valiant soldiers of the US Army fought six days here for what they deemed right on that blood-stained empty ground.

I was born on the Oregon Trail in western Nebraska. I grew to manhood in west-central Wyoming. As a boy I knew well the story of violent battles and collisions of will and purpose between the red man and white man in the American West long ago.

I learned the noble words "manifest destiny" distilled to their basics in history meant we, the white people, are the chosen race. In nineteenth century westward moving America, manifest destiny were noble words to white Americans, dangerous words if you were not. These words hid a sharp lingering prejudice that nonwhites were inferior. In short, do as whites say, get out of our way, or we will kill you.

I learned of the Lewis and Clark Expedition, 1804/06, and how a previously unknown tribe, the Nez Perce, and a Shoshone Indian woman, Sacagawea, saved the lost expedition from disaster. But white man's presence in the land of the Nez Perce on the Columbia Plateau in north-east Oregon and north-central Idaho, grew in the next seventy years from Lewis and Clark's first encounter with the Nez Perce. Violence came. Death approached the common. Soldiers and the federal government came to a once peaceful land to force their will and purpose on those who had come before them.

In June 1877 a few bands of Nez Perce refused a treaty forcing them to move to government reservations in today's north-central Idaho. These nontreaty Nez Perce then began a fateful flight to freedom. The Nez Perce War erupted between soldiers and nontreaty Nez Perce. It was indeed a war in purpose, intent, and in deaths.

The goal of the nontreaty Nez Perce was the vast plains to the east. But that changed. Betrayal from the Crow, once thought Nez Perce allies, forced the Nez Perce to look to unreachable freedom in Canada, or the "Old Woman's Country": named after Britain's reigning queen.

In their awe-inspiring and over 1,100-mile flight to freedom, the nontreaty Nez Perce outdistanced the best the Army could throw after them. They fought seven battles with the Army and several deadly skirmishes. The spirit of the nontreaty Nez Perce never broke, even after the last battle on that now sacred 100-acre field of honor in the north Montana prairie. Bear Paw Mountain was the last armed battle of the Nez Perce War: arguably the last Indian war in America. Here, as before, many lives were lost, and a legacy embellished beyond fact was born.

The flight of the nontreaty Nez Perce did not end at Bear Paw Mountain. For years to come, the US government forced them onto inhospitable lands of south-east Kansas and north-central Oklahoma.

If you know what lies before you on Snake Creek near Bear Paw Mountain, your vision and understanding changes from your eyes to your heart. On that October day, 2022, I began walking, hiking, driving, and photographing what is now called the Nez Perce National Historic Trail and beyond. Noted historians, the majority being white men, have published books, chronicles, and transcripts of the flight of the nontreaty Nez Perce on this trail. I will not repeat those chronicles here.

However, these accounts often gloss over or neglect the arduous nature of the over 1,100-mile trek across rivers, canyons, and mountains. A simple sentence of "they entered the canyon and crossed the river," belittles and mocks the jaw-dropping effort of the nontreaty Nez Perce.

Words tell, but photographs present. Few contemporary photographs exist of the flight of the nontreaty Nez Perce, which even today show the arduousness of the trek, the battlefields, rivers, mountains, canyons, and an expanse of endless prairies.

I hold a current remote pilot certificate from the US Federal Aviation Administration. I safely and expertly fly what the FAA terms a small, unmanned aircraft system or "drone." Modern aerial photographic technology using drones, brings rugged lands into better focus for the reader. These photographs show the nearly impossible effort, force of will, and resourcefulness the nontreaty Nez Perce needed to navigate majestic, aggressive, and potentially lethal land.

This work is not a history, but an earnest try in photographs to compliment or disprove that which authors have already written.

I looked at the situation, the terrain, land routes and battle grounds through my eyes: the eyes of a trained and combat experienced, now retired Army officer. My thoughts were many: Why did the nontreaty Nez Perce fail to keep a rear guard? Where were their reconnaissance scouts? How did they transit a seemingly impassable canyon channeling a fast-moving river? How could hundreds of cavalry soldiers first fail to outrun the Nez Perce, then finally catch the Nez Perce by surprise?

I concluded both the Army and the nontreaty Nez Perce violated two basic principles of war: Never get caught by surprise and never underestimate your adversary.

From personal battle experience in war, I know that battle reports, distances and details, are rarely accurate. War and battle are complex, fierce, and fast-moving dynamics, many times seen through frightened and exhausted eyes. Precise information can be fleeting, even among professionally trained and disciplined combatants.

No story told secondhand accurately reflects the emotion, the resolve, the fear, and the motive of the person who walked the trail before us, and about whose story we write. Judgement of men and women who walked the path we did not, and with whose priorities and prejudices we may disagree, is a siren's song writers, researchers, and historians alike take great risk in doing.

In my presentations and writings, I avoid the spontaneous attribution style of writing or the "he must have thought" manner of explanation (e.g., as Custer saw the hostile Indians surrounding him, he must have thought…). In most instances we do not know the thoughts of others. Presuming such psychic ability is not in anyone's best interest and does little to explain history.

My efforts for this book were not for notoriety or profit; I received no resource help, monetary help, or paid sabbaticals, from anywhere. My efforts were and are to help heal a wound I deemed needlessly and violently inflicted upon a proud and peaceful people caught in a shattering collision of cultures and purpose they did not create.

1

The People

In 1966 Everett S. Lee, University of Pennsylvania, wrote in his *Theory of Migration,* that people migrate to places because they are pushed somewhere, or they are pulled somewhere. Societal, environmental, economic, or cultural factors weigh heavily on this push-pull theory of how and why people live where they are. Mostly people move to survive.

America has had its indigenous peoples for eons. In America, as white people began a westward migration in the mid-to-late 1800s, they encountered more indigenous people. Many of these American Indians lived on the plains, valleys, mountains, and rivers of the vast American West. Something or someone pushed or pulled these American Indians to where white people encountered them. The pushing and pulling on the plains and beyond increased in earnest as white people came west over plains and mountains. Something pushed or pulled whites or American Indians, beit good or ill, to a perceived better life.

These indigenous people in the American West had names for themselves. Whites tried using English language and alphabet to translate these names into Anglican pronounceable words; not easy or accurate. The Nez Perce called themselves Niimiipuu (nee-mee-poo). Translated and pronounced as best any non-Nez Perce can, is "the people." The words "Nez Perce" (nezz purse) are mispronounced French words for "pierced nose." History is not sure where these words came from: Perhaps an English pronounced identifier probably from French-Canadian traders attributed to indigenous tribes of the region, including the Nez Perce. The name for the Nez Perce is inaccurate as they never pierced their noses, but these words and name have survived in history.

The first white men to encounter those called the Nez Perce was the 1804/06 Corps of Discovery, commonly called the Lewis and Clark Expedition. This was a US Army expedition with Meriwether Lewis and William Clark having officers's commissions and their men having enlisted men's ranks. President Jefferson chartered them to explore the newly acquired Louisiana Purchase, and also determine if a westward water route existed to the Pacific Ocean.

Led by Western Shoshone guide, Swooping Eagle or "Old Toby," Lewis and Clark labored west about 200 miles from today's Bitterroot Valley in Montana. They crossed a high and long east-west ridge called Lolo Trail. Lolo Trail was an old primitive through-way or trade route. Lewis and Clark did not create this trail. Indigenous peoples had been traveling this ageless trail or corridor long before Lewis and Clark. Saying that Lewis and Clark found the Nez Perce implies Lewis and Clark knew where the Corps of Discovery was going, when in fact they were essentially lost.

At Weippe (WE-ipe) Prairie at the west end of the Lolo Trail, Lewis and Clark came upon peaceful, resourceful, and strong people, the Nez Perce. Whether or not the Nez Perce were more surprised at seeing Lewis and Clark than the other way around makes a good argument in history: As in who discovered whom, and who were these hairy and perhaps smelly creatures walking off the Lolo Trail upright on two legs? Or in the case of York, Clark's man-slave, who or what was this creature having black skin and coarse buffalo hair.

In time Lewis and Clark spoke highly of the Nez Perce and

Nez Perce appaloosa horse in full regalia at the site of the Battle of Canyon Creek.

became in ceremony genuine friends with mutual respect. The Nez Perce provided resources, guides, and most importantly safety to take Lewis and Clark down the Clearwater River to near its confluence with the Columbia River. The Corps of Discovery (indeed America) owed their lives and knowledge of the region to the Nez Perce.

With the advent of the horse, indigenous peoples could better traverse the Lolo Trail corridor and into buffalo country of the Great Plains. Another Nez Perce name for themselves is translated as "the walking out people": meaning people who came from the mountains and onto the plains to hunt buffalo. The Nez Perce came to value horses. Horses meant stature, honor, and wealth. The Nez Perce homeland was excellent horse country, and the Nez Perce became superb horseman in riding and breeding. The Nez Perce had met with Plains Indians and partnered somewhat with the Crow. But in time the Nez Perce would find the Crow mercurial. Whatever cooperative existence between the two would vanish.

The Nez Perce were powerful people in a 10,000-square-mile region situated on the east-west trade and commerce land route from the Pacific Ocean to the Great Plains. The Nez Perce had the means to inflict great harm on those transiting Nez Perce land, but chose to greet others with diplomacy, tolerance, and respect.

Nez Perce lore says the Nez Perce did not migrate to their ancestral homeland on the Columbia Plateau of north-east Oregon, south-west Washington, and north-central Idaho. The Nez Perce say they were created there and settled in this peaceful and resource abundant scenic region. They called it home.

Tradition and lore say the Nez Perce began at today's Kamiah, Idaho, north of the confluence of the Middle Fork and the South Fork of the Clearwater River. An immense monster came to the valley and gorged itself on beings and animals living there. A being named "Coyote" outsmarted the monster, killed it, and used its flesh to create other bands of people. Using water from the Clearwater River, as Coyote cleansed the monster's blood from his hands, the blood-tainted water fell to the ground and created the Nez Perce.

The heart of this monster, the birthplace of the Nez Perce, exists south of Kamiah, Idaho. The heart, which Coyote excised from the monster, is a rounded short basaltic mound covered with wildflowers and other flora set off by itself in a lush pasture. The Clearwater River is a short distance west. Even with a highway nearby, Heart of the Monster remains a quiet and peaceful place. A place for reflection and thought.

The Nez Perce kept their deep-seated spirituality centered on the

Heart of the Monster, sacred to the Nez Perce near today's Kamiah, Idaho.

earth. They see the earth as the ultimate provider of life in all things. The earth is to be revered and cared for, not owned. Wayakin, a spiritual force unique to each member of a Nez Perce band, guided a Nez Perce in conducting their lives, making their decisions, and giving them insight to all things spiritual.

Even today this region of the Nez Perce west of Bitterroot Mountains and southeast of the Columbia River is somewhat isolated but remarkably beautiful with rich soil and abundant water. In prehistoric times a flood of soft basaltic lava covered this region. Four main rivers, the Salmon, Snake, Grand Ronde, and Clearwater, along with smaller streams and rivulets, have carved deep channels and canyons in the basalt. Environmental forces have rounded the basaltic hills and plateaus.

The Nez Perce still live in this vast region. It remains scenic but severe. The Nez Perce, like many American Indian tribes, were not a single unified people but rather a coalition or federation of similar bands having common ties and purposes. No single chief had dominion overall. The Nez Perce chose their leaders, headman or chief, by merit, demonstrated ability and leadership in peace and in war, not by hereditary title.

The Nez Perce had two headmen or chiefs named in English as "Joseph": Joseph the elder or "Old Joseph," and his son, Joseph the younger. Joseph the younger remains wrongfully in history's eye as some manner of supreme leader of all the Nez Perce. A capable leader, diplomat, and seeker of peace, Joseph the younger is indeed honored and impressive, but not omnipotent or supreme over all the Nez Perce.

Wallawa Lake in the Wallawa Valley
Joseph, Oregon

Wallawa Valley
Joseph, Oregon

Chief Joseph Canyon near Enterprise, Oregon. Chief Joseph and his band would winter deep in this canyon for shelter from storms and the cold.

Grand Ronde River canyon. Northern boundary of the Wallawa Band of Nez Perce.

Old Joseph's band of Nez Perce lived in the Wallawa Valley of north-east Oregon. Today only two out-of-the-way twisting and curving two-lane highways take you into the scenic and lush alpine-like Wallawa Valley. Today's valley is rich with alfalfa and hay fields. The bold Wallawa Mountains and Blue Mountains border the Wallawa Valley to the south and west. The long and thin deep blue Wallawa Lake, formed by glacier runoff and rock moraines, is south of today's Joseph, Oregon.

The Wallawa River enters the valley from the west and fills the Grand Ronde River. East of the valley the Imnaha (IM-nah-hah) River carves a deep crevice in the basalt creating a narrow, sheltered valley as the river flows northeast to the Snake River. Farther north the Grand Ronde River flows northeast to the Snake River. These rivers and valleys created informal boundaries among the several bands of the Nez Perce.

The rivers, fields, and valleys created a fertile and sheltered home for the Nez Perce. They were not a nomadic conquering people, they did not need to be. The Wallawa Valley and nearby lands provided them with all they needed to live and flourish.

But the missionaires, the black robes, came: First to honor the great commission from Jesus to "go ye therefore into all the world..." to save the eternal souls of the Nez Perce. Then in time, the whites came in abundance to take the Nez Perce land, their gold, and finally their lives. Manifest destiny loomed as dark clouds of an impending calamity.

Rattlesnake Hill north of Enterprise, Oregon, shows the many deep incised basaltic canyons of the Nez Perce country.

2

Upheaval

After the Lewis and Clark Expedition, fur trappers and traders came to the northwest. Among many diversed trade goods, traders brought guns. The Nez Perce noticed. In the 1830s and 1840s, white men came to Nez Perce country either to spread Christian salvation or to transit the region to the Oregon country or to the California gold fields. The white population grew in Nez Perce country. Cautiously the Nez Perce and the whites interacted, each laboring to make sense of each other.

But when two dissimilar cultures meet, turbulence and prejudice swell. Calm heads seek compromise and understanding; volatile and hotheaded minds seek subjugation and reprisals. A schism began among the Nez Perce: Do they seek partnership with the whites and their religion, or remain separate adhering to traditional Nez Perce ways and interpretation of life?

Whites began creating and expanding US government presence and control in the Northwest. In 1846 US and Canadian governments agreed upon the international border at the forty-nineth parallel of latitude. The US government created the Oregon and Washington territories in 1848 and 1853 respectively. Both territories included western edges of traditional Nez Perce homelands. With that, white man's laws, organization, and soldiers came to Nez Perce country.

In May 1855 Washington territorial governor, Issac Stevens, an impetuous former Army officer and Civil War veteran, wanted a lasting agreement with the tribes in the territory. Part of that agreement was identifying Nez Perce land, which in some manner Stevens wanted to control. Indian lands, however, did not begin and end precisely with a surveyor's chain and transit, but rather a consensus among nomadic people as to where their land began and ended. Stevens invited tribes on the Columbia plateau to a meeting. The tribes arrived in force.

Here Stevens explained the government's position that the

government wanted the tribes on reservations, along with other stipulations. Stevens looked to have a signed treaty with the tribes, including the Nez Perce. Stevens looked to anyone supposedly in authority to speak for the entire Nez Perce people.

But the tribes did not have a governing structure with one person in charge. The Nez Perce had several bands or divisions with similar characteristics, but different leaders. Separate bands within a tribe looked to consensus rather than a single person ruling structure. The separate leaders could not commit the other bands to anything. This apparently meant little to Stevens. He wanted a treaty and would work with anyone to get it.

Stevens found a Nez Perce that Stevens—wrongfully—concluded had over-arching authority: a Christian Nez Perce with the Anglican name, ironically, of Lawyer. Lawyer was some manner of accepted voice of the northern Nez Perce, friendly to the whites and Stevens. Lawyer signed the treaty. That satisfied Stevens, but not all Nez Perce.

Looking Glass
(National Park Service)

Looking Glass, a Nez Perce elder and honored chief, a true chief, arrived late almost by happenstance at the treaty meeting. Looking Glass railed against the treaty as giving away Nez Perce land. But it was too late. The deed was done. The treaty allowed more whites to come.

For a generation from the mid-1850s to the mid-1870s this so-called treaty period gave rise to a heritage of grievance, instability, and insecurity. The schism between the northern Christian Nez Perce who supported the 1855 treaty, and five bands of the southern non-Christian Nez Perce who did not, (nontreaty Nez Perce) widened to a terrible gap with deadly consequences soon to come.

Gold—white man's treasure, red man's curse. In late 1860 prospectors discovered gold in the north-east part of Nez Perce country near today's Pierce, Clearwater County, Idaho. Miners traveled to the region along the banks of the Clearwater River, then crossed the river and climbed up and out of the scenic canyon carved by relentless and wide Clearwater River. Thousands of whites came and ignored treaty stipulations. Hostilities grew between Nez Perce and whites. Prospectors did not come to the remote Wallawa Valley, but in time ranchers and white settlers soon learned of the valley and its riches in land and water.

By 1863 white authorities looked to avoid war. They convened another treaty negotiation meeting at Fort Lapwai on the reservation. This meeting reduced Nez Perce land further and favored northern and Christianized Nez Perce at the reservaton. More whites came to Nez Perce land and stayed, many times without authorization. Random killings began. Justice favored the whites, but diminished the Nez Perce.

General Oliver Otis Howard
Right arm missing
(NARA)

To maintain order, in 1876 the Army arrived with Brigadier General Oliver Otis Howard, officer commanding, Military Department of Columbia. Howard had lost an arm in the Civil War and was former head of the Freedman's Bureau, formed to assist freed slaves. He was known as a "Christian general" with deep cords of piety and liberal thought framing his actions. Regardless of Howard's philosophical foundations, storm clouds and lightning, once distant, now inexorably rolled toward him and the nontreaty Nez Perce.

Old Joseph died in 1871. Before Old Joseph died, he cautioned his son, Young Joseph, not to give up Nez Perce ancestral lands. Old Joseph's people buried him near today's Wallawa Lake. Young Chief Joseph then became the headman of the Wallawa band of nontreaty Nez Perce. Young Chief Joseph had become of age and made a commanding but diplomatic presence at a November 1876 meeting with Howard and other commissioners at Fort Lapwai. But the government's position held: All Nez Perce, including nontreaty Nez Perce, had to move to the reservation by mid-June.

In May 1877 nontreaty Nez Perce tried in vain to get Howard and Bureau of Indian Affairs agent John Montieth to relax the government's position that all Nez Perce had to move to the reservation within thirty days. This was a daunting mandate to nontreaty Nez Perce as they lived mostly in the Wallawa Valley with fierce rivers and savage canyons to

Chief Joseph the Younger (Smithsonian Institute)

cross to reach the reservation to the north along the Clearwater River.

Gathering belongings and what livestock and horses they could, young Chief Joseph and his band of nontreaty Nez Perce began the arduous trek from the Wallawa Valley north to the reservation. The nontreaty Nez Perce left little account of this, but walking and visiting this land today instantly shows the herculean effort needed to transit here. Among the many challenges in this geography was descending over 2,000 feet in elevation from the Wallawa Valley plateau to the Imnaha River. This river carved another narrow canyon toward the Snake River.

Climbing steep canyon walls out of the Imnaha River valley, the nontreaty Nez Perce crossed a low point in a ridge west of the Snake River called Lone Pine Saddle. Descending over 1,500 feet to the Snake River, nontreaty Nez Perce faced the fast-flowing Snake River. Spring rains and snow melt had filled the river to foaming and treacherous levels. The Nez Perce were in today's Hells Canyon, the deepest river gorge and canyon in North America—deeper than the Grand Canyon.

Imnaha River Valley. Nez Perce bands lived in this valley northeast of the Wallawa Valley.

Toward the end of May 1877, Chief Joseph and his band of nontreaty Nez Perce crossed the formidable Snake River at today's Dug Bar Crossing. Women, children, and the elderly crossed on horsehide rafts pulled by horses. Many horses and livestock drowned in the crossing. The nontreaty Nez Perce abandoned many of their livestock on the river's west bank. These Nez Perce would face more rivers and canyons in the coming months. Unknown to them, many would never again see their homeland in the Wallawa Valley.

Continuing northeast from the Snake River, Chief Joseph's band of people and thousands of horses, labored again across a high divide between the Snake and Salmon rivers. Crossing the Salmon River, also at flood stage, in early June 1877 they reached an open prairie west of today's Grangeville, Idaho, called Camas Prairie. This prairie had a natural lake carved in antiquity called Tolo Lake.

Tolo Lake was an ancient meeting ground for the nontreaty Nez Perce. The 40-acre lake today sits as a round recreational pool on grassy prairie at the head of Rock Creek and Rocky Canyon. What formed this lake is uncertain, perhaps artesian waters. No stream appears to feed or empty it. (When the author visited Tolo Lake in June 2024, campers and families paddled on the lake, splashing and having a good time.)

Five nontreaty Nez Perce bands camped together at Tolo Lake that early June 1877. History says about 600 nontreaty Nez Perce gathered, including women, children, and the elderly. How many were men of fighting trim is not known, probably less than half.

The bands rested. For food, women dug and gathered camas bulbs. The light blue camas flower is native to North America. The camas bulb is a food source from its edible bulb. However, between festivities uneasiness lingered. Bands conversed as to what to do next. Then something happened. A small spark lit a fuse racing toward a powerful keg of upheaval. Whatever constraints holding that keg instantly snapped.

Younger spirited nontreaty Nez Perce warriors at or near Tolo Lake, perhaps fueled by liquor, sought revenge on white settlers for long simmering grievances and white killings of Nez Perce. Nontreaty Nez Perce killings of whites and rampages began. Chief Joseph, away from the Tolo Lake gathering, learned of the killings when he returned.

The nontreaty Nez Perce attacks during the second week of June 1877 around Camas Prairie caused the Army at Fort Lapwai to mobilize and move upon the nontreaty Nez Perce. Chief Joseph, always seeking peace, made a heart-rendering decision to move his band for safety to the camp of White Bird at the southern end of a canyon about ten miles south of Tolo Lake: a canyon soon to flow red with blood.

Tolo Lake on Camas Prairie near today's Grangeville, Idaho. View west toward Rocky Creek and Rock Canyon. Both lead to the Salmon River. The canyon was a natural path for the Nez Perce coming east from the Salmon River and toward Camas Prairie.

Tolo Lake looking east toward today's Grangeville, Idaho.

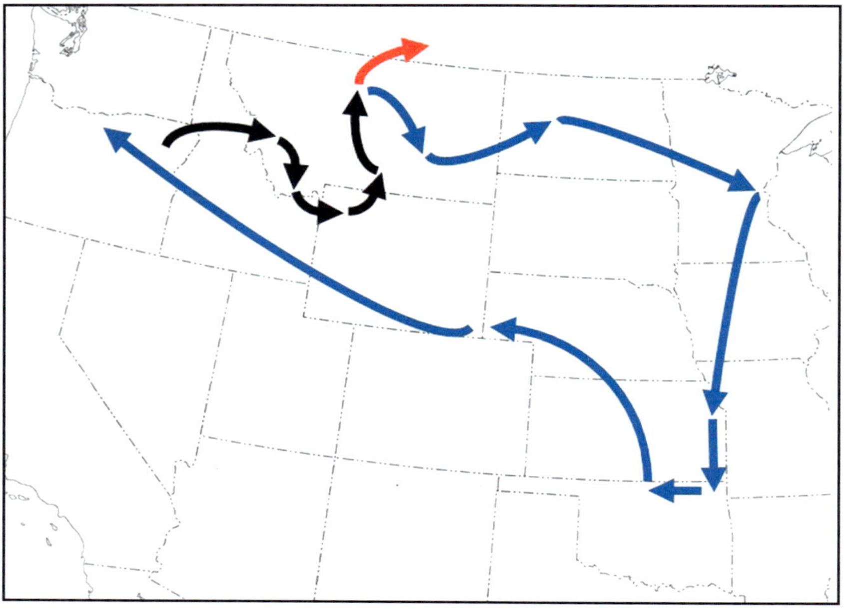

Flight of the nontreaty Nez Perce

Black arrows show the war route, June to October, 1877.
Red arrow shows White Bird's escape to Canada after Bear Paw Mountain.
Blue arrows show the route of the exiled nontreaty Nez Perce from 1877 to 1885.

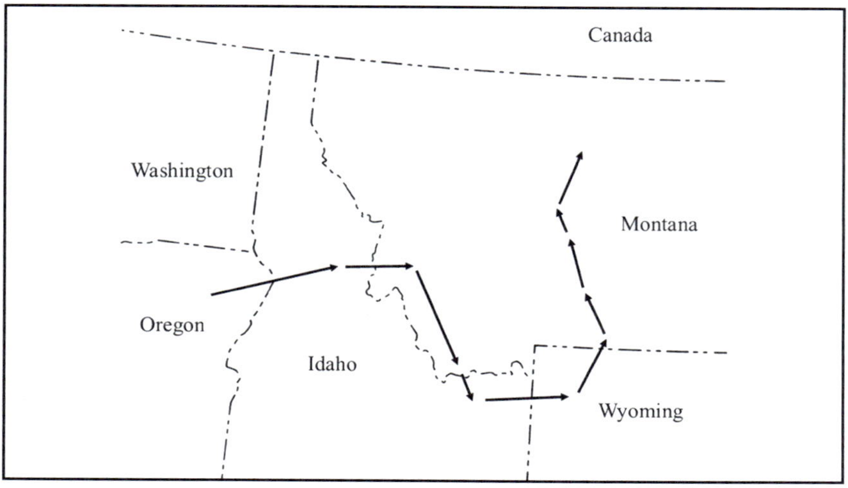

Flight of the nontreaty Nez Perce, June to October, 1877.

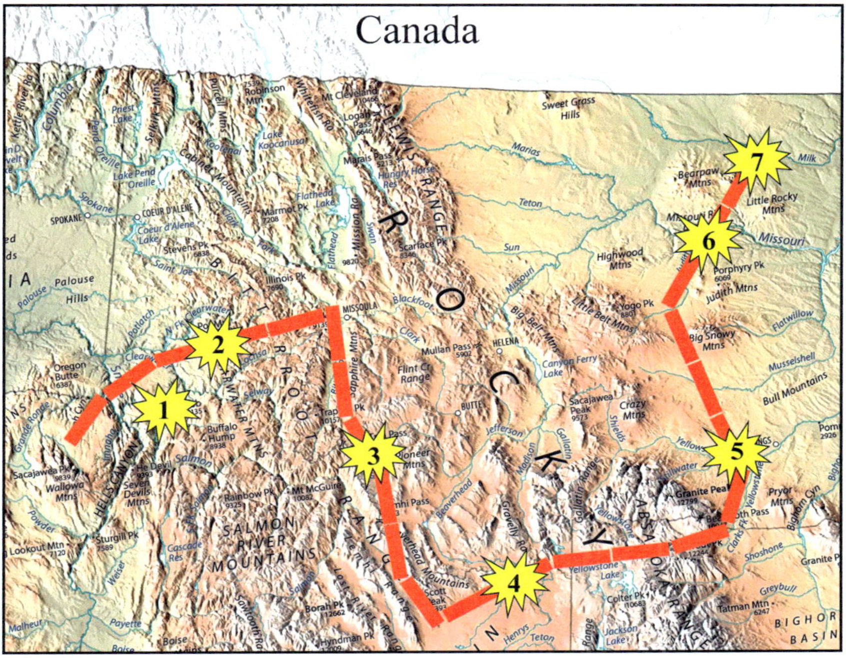

US Geological Survey topographical relief map

Major battles of the Nez Perce War, 1877, along the route of the nontreaty Nez Perce flight to Canada.

1. White Bird Canyon, Idaho: June 17
2. Clearwater, Idaho: July 11, 12
3. Big Hole, Montana: August 8, 9, 10
4. Camas Meadows, Idaho: August 20
5. Canyon Creek, Montana: September 13
6. Cow Island Landing, Montana: September 23, 24
7. Bear Paw Mountain, Montana: September 30 to October 5

3

War Begins

General Howard heard of the killings and rampages near Tolo Lake. He had little time to send soldiers south to restore order and protect white settlers in that area. Howard ordered about 100 soldiers and officers to depart Fort Lapwai and move about sixty miles south to where civilians reported hostilities. This force reached Camas Prairie and Grangeville on June 16, 1877. Records read this was a well-equipped force led by experienced Army officers. The force included eleven civilian volunteers and twelve Nez Perce scouts who allied with Howard.

White Bird was chief of his Nez Perce band. They settled along the right bank (looking downriver) of the Salmon River southwest of today's Grangeville, Idaho, and south of Tolo Lake. White Bird had his camp along White Bird Creek upstream from its mouth on the Salmon River. A steep canyon, the watershed of White Bird Creek, rises from the creek and extends north. From its southern lower end at the creek to its upper northern end, White Bird Canyon rises about 2,000 feet. A steep-sided ridge with two large hills divides the canyon north-to-south. Feeder seasonal creeks have carved deep rivulets and gullies in the canyon walls.

On Sunday morning, June 17, Captain David L. Perry, a seasoned officer and leader, approached White Bird Canyon from the north. Relentless marches in bad weather had exhausted his men and their horses. Perry saw White Bird's camp at the lower end of the canyon. Nez Perce accounts read the Nez Perce learned of the soldiers's approach, and sent a small party of Nez Perce under a flag of truce to determine the soldiers's intention. Nez Perce accounts also emphatically state that a civilian white man (possibly Army scout Arthur "Ad" Chapman) opened fire on the Nez Perce without provocation. The battle began in earnest.

Canyons are a curse to an attacking force, and a blessing to a defending force. White Bird Canyon's hills, gullies, and lack of cover divided and stalled the Army's attack. Nez Perce warriors, estimated at between sixty and seventy in number, pressed the soldiers on the Army's flanks and gained an advantage. Long range Nez Perce rifle fire cut down Perry's men.

At that time bugle calls piercing the crescendo and din of combat on battlefields were the primary method of command and control of soldiers. At White Bird Canyon long-range gunfire early on killed Perry's two buglers. The battle began unraveling for the Army.

The Nez Perce were killing Perry's men, and most importantly Perry was losing command and control. His men, many were inexperienced junior enlisted men and a few unreliable civilian volunteers, began panicking and withdrawing north up the canyon. Perry tried his best to reform his battle lines, but failed. The Nez Perce routed Perry's force from the canyon and drove them miles beyond.

An oddity in battle is a soldier's sense of time. In combat, time seems to slow. Minutes seem like hours. An Army survivor of White Bird Canyon later said the battle lasted about fourteen hours, when in fact it lasted not quite three hours.

Thirty-four soldiers died in battle at White Bird Canyon. No Nez Perce were killed-in-action there. Ten days after the battle the Army hastily buried the dead soldiers where they fell. Unlike many instances with Plains Indians, the Nez Perce regularly did not scalp or otherwise mutilate the dead. In fact, during the Nez Perce War, no instance was recorded where the Nez Perce violated the dead of their adversaries.

The Army had learned the Nez Perce were formidable, seasoned warriors, skilled marksman, and expert horsemen. The Nez Perce would fight with discipline and courage. The Nez Perce War began.

To the victor belong the spoils. Nez Perce warriors recovered quantities of modern-for-their-day Army arms and ammunition from the battlefield. These weapons served well the Nez Perce in days to come.

As with the Battle of the Little Bighorn (Custer's Last Stand) the year before, the inevitable finger-pointing began as to what happened to the Army at White Bird Canyon. In short, the Army underestimated their adversary on terrain and ground the Army knew little about. A disciplined and determined force outfought the Army. The Army inflated casualties and forces involved. The Army was not going to officially say it had been soundly whipped by Indians fewer in number.

Unlike soldiers at White Bird Canyon, the Nez Perce had a definitive and personal reason to fight. They fought to protect their homeland and their families.

White Bird Canyon and White Bird Creek. View south toward the town of White Bird, Idaho, in upper center. Chief White Bird had his village along White Bird creek at left. The Army attacked the village from lower right to center in the photograph. Today's highway 95 at right coming down a steep incline.

White Bird Canyon. View north from near White Bird's village. The Army attacked from the north but rugged terrain shown (two hills and rocky ridge) impeded and divided the Army's attack. Nez Perce warriors counter-attacked around the hills and drove the soldiers back up the canyon.

White Bird Canyon and White Bird Creek. View south toward the town of White Bird, Idaho, in upper center. Chief White Bird had his village along White Bird creek. The Army attacked the village going downhill around the two hills shown. Nez Perce warriors drove the soldiers back from the two hills and up the canyon (north).

White Bird Canyon. View north. The Army attacked from the north toward the two hills in center. Nez Perce warriors counterattacked around the hills and routed the soldiers back up the canyon, upper right in photograph. Nez Perce warriors pursued the soldiers beyond the canyon for several miles.

Howard learned of the Army's defeat and called for reinforcements. This was critical as the Army could replenish its losses in manpower and equipment, the Nez Perce could not. Howard and about 400 soldiers moved south from Fort Lapwai. Howard assumed direct command of the force.

For the next three and a half months he would chase the illusive and resourceful Nez Perce. Howard knew of Chief Joseph and Joseph's eloquence in speech. Howard began to infer—wrongfully—that Joseph was the headman of all nontreaty Nez Perce. Howard never caught them. Not so for an aggressive hell-bent-for-leather Army commander based along the distant Yellowstone River, Colonel Nelson A. Miles.

For the next month after White Bird Canyon a confused hide-and-seek, cat-and-mouse situation began with the Army and nontreaty Nez Perce, including Chief Joseph. The Nez Perce considered their options. Chief Joseph, himself a new father with a baby girl, pondered a peaceful solution, perhaps maybe a peace delegation of some manner. But getting to safety away from the Army was Nez Perce priority.

Rivers served well the Nez Perce. Southwest of White Bird Canyon, the Snake and Salmon rivers run north parallel to each other. At their nearest the rivers are about eight miles apart. Between them is rugged country, with high peaks, steep valleys, and canyons. A perfect hiding place the Army knew nothing about. The Nez Perce knew this country and decided to go there. The Army failed, again, to determine what manner of terrain or ground the Army would encounter in pursuing the Nez Perce across the Salmon River.

Rivers in Nez Perce country still ran high in mid-June. From White Bird Canyon the Nez Perce moved about ten miles south to a sharp bend in the Salmon River called Horseshoe Bend. Crossing the turbulent river here from east to west on June 19, the Nez Perce moved west into the rugged country between the Salmon and Snake rivers.

After stopping in Grangeville to visit wounded soldiers and at White Bird Canyon to bury dead soldiers, Howard pursued the Nez Perce south to Horseshoe Bend. But he and his men were woefully deficient and lacking engineering skills to ford fast-moving water. Howard needed three days to cross east-to-west the turbulent and swollen Salmon River. His force crossed by July 1, but by that time the Nez Perce had moved north leaving the Army behind.

Chief Looking Glass's ancestral home was near the confluence of the Middle and South forks of the Clearwater River southeast of today's Kooskia, Idaho. Looking Glass and his band had no part in the battle of White Bird Canyon. Howard had heard that Looking Glass probably was

The site of Looking Glass's village near today's Kooskia National Fish Hatchery; Kooskia, Idaho. View south. Soldiers attacked the village from the high ground west (right in photograph).

The site of Looking Glass's village near today's Kooskia National Fish Hatchery; Kooskia, Idaho. View south. Soldiers attacked the village from the high ground west (right in photograph).

not a threat. But Howard began believing reports that Looking Glass had in some manner supported or would join Chief Joseph. Howard had a change of heart and wanted Looking Glass arrested.

Howard ordered Captain Steven G. Whipple with a force of over sixty soldiers and twenty volunteers to attack Looking Glass's village and arrest the chief. Going east to the village, again, proved arduous due to unrelenting hilly terrain, steep incised canyons, and the South Fork of the Clearwater River. Looking Glass's village was along Clear Creek, which empties into the Middle Fork of the Clearwater. (The village was near today's Kooskia National Fish Hatchery.)

As with many battle reports, accounts vary as to who or what started the gunfire. On July 1, Captain Whipple arrived on a hill or mountain 550 feet higher in elevation west of Looking Glass's village. Whipple reported he (Whipple) had announced his presence and gave Looking Glass an opportunity to surrender. Nez Perce accounts say otherwise. They say the Nez Perce wanted a parlay of sorts to ensure Whipple knew Looking Glass did not want war. As Whipple and a small contingent approached Looking Glass's village, someone on the hill to the west fired a shot. Looking Glass and his people scattered.

Looking Glass's scattered warriors were not arrayed to defend the village. Soldiers burned Nez Perce lodges and trampled their garden plots. Nez Perce casualties are uncertain, perhaps three or four killed, with others wounded. Whipple and his soldiers suffered no casualties. They did not capture Looking Glass. However, of great loss to Looking Glass was soldiers capturing hundreds of Nez Perce horses.

Whipple failed in his mission to arrest Looking Glass. But the attack cemented Looking Glass's resolve to resist soldiers and the Army. Looking Glass was a capable leader and knowledgeable of buffalo country far to the east. He would be a great asset in the weeks to come. He had also fought as an ally of the Crow.

Howard was moving north up the left bank (looking downriver) of the Salmon River, while the Nez Perce had crossed back over the Salmon River near today's Billy Creek at a place once called Craig's Ferry: about fifteen miles west of today's Cottonwood, Idaho. The Nez Perce then headed east across the wide-open Camas Prairie.

About twenty miles west of Looking Glass's village on the rolling plains north of the Salmon River, Benjamin and Jennie Norton had acquired a station or ranch of sorts on the road from Lewiston to the town of Mount Idaho. But during the June 14/16 Nez Perce raids and killings, the Nez Perce had either killed or driven off the Nortons and others near the settlement.

After attacking Looking Glass's village, Captain Whipple retired to the Norton settlement (today's Cottonwood, Idaho). Army scouts reported Nez Perce in the area with one scout wounded and possibly left behind. Whipple ordered Lieutenant Sevier M. Rains to take ten hand-picked men with one scout to find the man left behind and determine the strength of the Nez Perce. Camas Prairie is about 750 square miles of fine horse country, and Rains and his men rode into mounted Nez Perce warriors. The warriors destroyed Rains and his men. Whipple came from behind Rains but did not pursue the Nez Perce.

About noon on the morning of July 5, along the road near the Norton settlement, the Nez Perce attacked a group of seventeen civilian volunteers led by Darius B. Randall. Randall, a Civil War veteran, apparently wanted to attack the Nez Perce in the Cottonwood area. But the mounted Nez Perce killed Randall and others. Whipple did not counterattack. The Nez Perce then followed their familiar trails east across Camas Prairie to the forks of the Clearwater River.

The disjointed Army and volunteer effort did little to stop the Nez Perce. What it did was to divert the Army's attention from the Nez Perce gathering at the mouth of Cottonwood Creek along the south fork of the Clearwater River. Less than a week later the Army and Nez Perce would meet again in force.

Camas Prairie near today's Cottonwood, Idaho. View east from highway 95 south of Cottonwood. Nez Perce warriors attacked and destroyed Lieutenant Sevier M. Rains and later Darius B. Randall on this prairie. Randall and his men fought the Nez Perce near the farm house shown.

Camas Prairie near today's Cottonwood, Idaho. Memorial to Darius Randall and his men, many of whom died fighting Nez Perce warriors on July 5, 1877. Randall and his men fought the Nez Perce near the farm house in the far ground. Memorial is along highway 95 south of Cottonwood, Idaho. Memorial was erected July 5, 1931.

The vast Camas Prairie near today's Cottonwood, Idaho. Site of many deadly skirmishes between Nez Perce warriors and white settlers.

4

Clearwater And East

The first of July 1877 saw confusion and disarray in the Army's mission to subdue, capture, or kill the nontreaty Nez Perce in the region of the rugged breaks and canyons of the Salmon River and in Camas Prairie. Howard lacked manpower, experience, and skill to corral the Nez Perce in those regions. While the Nez Perce had crossed west-to-east back over the Salmon River near defunct Craig's Ferry, the river had defeated Howard again. He withdrew southeast about twenty-five miles over demanding terrain to his earlier crossing near the mouth of White Bird Creek south of White Bird Canyon. Howard needed two days to get his force back across the swollen Salmon River.

By Monday, July 9, he had moved north to Grangeville through the bloodied ground of White Bird Canyon. His men and horses were weary and worn out. The Nez Perce were also in disarray. Who was in charge and where should they go?

After recrossing the Salmon River, four bands of Nez Perce, (i.e., Joseph, Ollokott, White Bird, and Toohoolhoolzote) began moving about thirty miles east toward the forks of the Clearwater River and Looking Glass's village. As the bands moved east, Nez Perce raiders killed several white settlers and burned settlements. The bands followed Cottonwood Creek drainage and canyon. The creek flows into the South Fork of the Clearwater River near today's Stites, Idaho.

The bands set camp along both sides of the South Fork, but mostly on the west side of the river as it had more room. Looking Glass's people soon arrived. This increased the nontreaty Nez Perce strength to about 750 people, but not all were warriors.

A civilian volunteer force led by "Colonel" Edward McConville, began following the Nez Perce across Camas Prairie. On July 9, McConville and his men, having followed Cottonwood Creek, discovered the Nez Perce camp along the South Fork. McConville sent messengers

Battle of the Clearwater. View north towards today's Stites, Idaho. South Fork of the north-flowing Clearwater River in middle. Cottonwood Creek empties into the South Fork at lower left. Nez Perce village was on both sides of the South Fork. The Army attacked the village from the high ground to the east (right in photograph).

to Howard and then waited on a plateau north of the Nez Perce camp. The Nez Perce discovered McConville's men on the plateau, later called "Misery Hill", and besieged them for two days before withdrawing.

Howard with a force of about 350 men was south of the Nez Perce camp when he learned of the Nez Perce along the South Fork. He moved northeast to get on high ground. On Wednesday, July 11, Howard was on a high, grassy, treeless basaltic plateau east of the Nez Perce camp and south of the destroyed Looking Glass village. Howard saw the Nez Perce village below him along Cottonwood Creek and the South Fork of the Clearwater River. He positioned his artillery on high ground about 900 feet above and northeast of the village and opened fire around noon. The Battle of the Clearwater had begun.

The two-day Battle of the Clearwater was one of the few, if not the only battle in the Nez Perce War or the Plains Indian Wars, where demanding terrain forced both sides to form battle strategies and maneuver on their adversary. This was not a characteristic lightning fast, strike and run, mounted horse battle known with soldiers and Plains Indians throughout the American West of that era.

But in battle the Nez Perce fought as individuals or small groups who saw and agreed upon a common purpose and a common leader. Their leader was situational and consensus based. At the Clearwater soldiers and warriors fought with courage and skill. But as the battle

Battle of the Clearwater. View southwest toward today's Stites, Idaho. The Nez Perce had their village along the South Fork of the Clearwater River in middle. Howard placed his artillery on the flats in lower right to bombard the village without effect. Howard and the remainder of his soldiers were on the high ground east of the village (left in photograph).

Battle of the Clearwater. View west from the high ground east of the Nez Perce village along the South Fork of the Clearwater River. Cottonwood Canyon in far ground (upper center in photograph). Howard formed a battle line across the meadow (middle in photograph). Nez Perce warriors attacked up the draws north and south of Howard's position (far right and far left in photograph). On the last day of the battle, Howard ordered a charge down the left draw and drove the Nez Perce from the field.

Battle of the Clearwater. View east toward the high ground above the Nez Perce village on the South Fork of the Clearwater River. Howard had his soldiers on the open area in upper center of photograph. Nez Perce warriors attacked up the valleys north and south (left and right in photograph).

progressed and entered a second day, that common purpose among the Nez Perce waivered as did their leadership. In the life of the Nez Perce, Shoshone, and other Plains Indians, Meriwether Lewis wrote, "Each individual man is his own sovereign master, and acts from the dictates of his own mind."

The soldiers followed orders. They fought as units commanded by a single over-arching officer, the Nez Perce did not. The Army was an immense regimented structure with vast resources of men and materiel. In the Nez Perce War the Army would call upon this vastness while the Nez Perce had none. The cause of the nontreaty Nez Perce, while valiant and inspiring in history, was in fact doomed.

At the Clearwater, the Army held fast on the high ground: This was a plateau about a mile wide north-to-south, and a mile and a half long, east-to-west. The Nez Perce had to maneuver up precipitous breaks north and south of this plateau to counterattack the Army. In reverse, the Nez Perce commanded these ravines off the plateau and down to the South Fork and the Nez Perce village.

This was a dismounted infantry battle. Cavalry was largely ineffective in this demanding arduous country given constraints of the high ground where Howard formed his men.

Battle of the Clearwater. View west from the high ground east of the Nez Perce village along the South Fork of the Clearwater River. Cottonwood Canyon in far ground (upper center in photograph). Howard formed a battle line with his troops across the meadow (right in photograph). Nez Perce warriors attacked up the draw (left in photograph). On the last day of the battle, Howard ordered a charge down the left draw and drove the Nez Perce from the field.

Battle of the Clearwater. View west from the high ground east of the Nez Perce village along the South Fork of the Clearwater River. Cottonwood Canyon in upper left in photograph. Howard placed his artillery on the meadow northeast of the Nez Perce village (upper right in photograph). Nez Perce warriors in a two-prong attack came up the draw shown in middle along the road.

Battle of the Clearwater. View north on the high ground east of the Nez Perce village. Howard positioned his soldiers in a battle line across this meadow. Nez Perce warriors attacked from left in photograph. After the battle, Howard buried his dead somewhere in this meadow. As of this writing it is not known the disposition of the bodies of the soldiers killed-in-action on this meadow.

Howard's artillery initially proved useless as the cannons were positioned high above and distant from the Nez Perce camp. But these cannons with black-powder propellant charges are deafening. They belch fire and smoke. These weapons terrify anyone not familiar with them, including the Nez Perce in the valley below. Howard moved his artillery south on the plateau east of the river, but again his cannons and Gatling guns proved ineffective.

Nez Perce warriors, initially taken by surprise, began moving east and up the steep ravines north and south of Howard's position on the plateau. The main Nez Perce firing positions were in these ravines near the top of the plateau. Near the top, Nez Perce warriors created rock barriers and poured extensive fire into the soldiers. Warriors did concentrate fire on one cannon killing most of its crew.

Toward late afternoon the battle subsided and settled into desultory sniping. Nez Perce gunfire forced soldiers to dig rifle pits and stay low. Lack of drinking water became critical for the soldiers. The Nez Perce had access to water below, while the soldiers did not. A natural spring was atop the plateau but within rifle fire distance of the Nez Perce. Nez Perce gunfire kept the soldiers from the spring until dark.

Night came and both sides kept their positions. Many Nez Perce warriors returned to their village along Cottonwood Creek. Dissention and discord began among Nez Perce warriors. Many thought the fight was not worth further struggle. The Nez Perce also did not have a single headman in charge at this fight. Nez Perce accounts state that Chief Joseph was not in the fight, but rather at the camp organizing the people and saving their horse herd. Although the Nez Perce gave a disciplined fight, they still did not function as a unified force.

At daybreak Thursday, July 12, the battle continued with no side gaining a tactical advantage. By afternoon Howard planned a coordinated assault on the Nez Perce positions guarding the ravines down from the plateau. At 3:00 p.m. Howard ordered a charge. Soldiers massed along the steep south ravine and charged. The onrush drove the Nez Perce down the ravine and across the South Fork of the Clearwater River. The Nez Perce abandoned their camp and continued west along Cottonwood Creek. The soldiers did not cross the South Fork in force.

The battle ended with soldiers and civilians looting then burning the abandoned Nez Perce village.

Howard had a solid chance to pursue the retreating Nez Perce and finish the job, but Howard inexplicably did not follow-up his battlefield success. For a Civil War combat veteran and leader not to follow-up when the enemy was on the run was tantamount to dereliction of duty and a serious battlefield mistake. Others, including newspapers, began criticizing Howard for not being aggressive and allowing a smaller force to wage battle and smite the Army and escape.

Today's Clearwater battlefield, largely a hay field. View west toward distant Cottonwood Creek canyon. Nez Perce warriors attacked up the draw at left.

Clearwater River forming the western boundary of Weippe Prairie.

West end of Weippe Prairie overlooking Clearwater River canyon. Nez Perce warriors fired upon Howard's soldiers as the soldiers climbed from the canyon to the prairie.

Howard reported Army casualties at the Clearwater were twelve men killed, twenty-seven wounded, and one missing. Howard buried the dead soldiers in a trench on the plateau east of the South Fork. Nez Perce said their casualties were four killed and about six wounded. Howard called the battle a victory for the Army, but the Army had not defeated the Nez Perce and ended the war there. The Nez Perce also had learned a lesson at the Clearwater: The Army was becoming a worthy adversary.

The nontreaty Nez Perce headed west from their camp on Cottonwood Creek, then turned north to Kamiah. Kamiah was on the Nez Perce reservation. Nez Perce in Kamiah were treaty Indians and did not cooperate with the nontreaty bands coming north.

The nontreaty bands crossed the river and began the 2,000 feet in elevation climb from the river east to Weippe Prairie. This was the same prairie where the Nez Perce saved Lewis and Clark over seventy years before. (Even today, this is a steep twisting and turning road to Weippe, Idaho.)

Howard's troops saw the Nez Perce on the steep canyon walls east of the river. Rearguard Nez Perce warriors waited and opened fire, driving the soldiers back.

Howard then received a message that Chief Joseph wanted a parlay or meeting to discuss a possible surrender. Howard seemed cautiously optimistic that the message was valid. But Joseph never appeared. Who did indeed appear was about fifty Nez Perce led by Red Heart. He and his band were returning from hunting on the plains. They were not involved in the fighting. They were peaceful people and had met the nontreaty Nez Perce. Red Heart decided not to join them.

Red Heart and his band met bad luck. Howard arrested Red Heart. In July heat soldiers marched Red Heart and his band sixty miles to jail at Fort Lapawi, and then to Fort Vancouver, Washington, where they would spend months in captivity for something they did not do.

On Monday, July 16, soldiers crossed the Clearwater River to find the Nez Perce gone. The Nez Perce had held a council at Weippe Prairie to determine what to do. The Nez Perce, with debate and worry, decided to move east across Lolo Trail: the ancient trading corridor, and into Montana's Bitterroot Valley. Hopefully they would find friendly whites and other tribes they had allied with. Also, maybe in Montana they could leave the Army behind.

But the Army knew no regional boundaries. The Army would pursue the Nez Perce relentlessly wherever the Nez Perce went.

5

Lolo Trail And The Bitterroots

The Corps of Discovery, or the Lewis and Clark Expedition, recorded the best and most accurate description of its day in crossing the Bitterroot Mountains and navigating Lolo Trail. Bitterroot Mountains and Lolo Trail remain an arduous, unsettled barrier of severe but magnificent country. Even today it is passable but not without resolve and purpose. These mountains and treacherous canyons with fast-moving rivers nearly swallowed the Corps of Discovery.

On Monday, September 9, 1805, the Corps of Discovery, having entered Montana's Bitterroot Valley from the south, camped at the confluence of today's Lolo Creek and north-flowing Bitterroot River: a place called "Traveler's Rest" southwest of today's Missoula,

View west. Site of Traveler's Rest (middle of photograph) south of today's Missoula, Montana.

Montana. Looking west, Lewis and Clark saw formidable snow-covered mountains. Lewis and Clark recorded the initial part of the trail east of the continental divide as tolerable. The expedition crossed today's Lolo Pass and descended into a fine meadow now called Packer Meadows.

The expedition then began following west-flowing Lochsa River. (Lewis and Clark called this river "Kooskooske River.") The Lochsa would soon meet the Selway River. These two formed the wide, scenic, and fast-moving Middle Fork of the Clearwater River. These rivers cut steep channels in the mountains with rocky banks not suitable for overland passage by foot. In places an intrepid person could float the rivers. (Commercial white-water float trips still do.) But snow began falling, and the expedition did not have time to halt, build wooden canoes, and try their luck floating an unknown river with foaming rapids.

Along the Lochsa River the expedition's guide, "Old Toby," got lost. The expedition had to move north up steep, aggressive ground covered with rocks and fallen timber to an east-west ridge. (This ridge is north of and parallels today's highway 12.) This ridge is where the trail, even today, goes between Weippe Prairie and the Bitterroot Valley. On this rocky ridge the expedition was in desperate condition: Starvation, exposure, and declining morale enveloped the men. Disaster loomed.

Confluence of the Selway (left) and Lochsa (lower right) rivers. Together they form the Clearwater River (right center). View west, downriver of the Clearwater.

Overleaf. Packer Meadows looking east.

Packer Meadows with camas plants in full bloom.

Camas plant in full bloom. The bulbar plant has nutricious blubs, which the Nez Perce harvested and used as food.

Today's Lolo Trail Motorway, which follows the old Lolo Trail. This is an all-wheel drive, high-center vehicle trail only in dry weather.

Lolo Trail foot path

View south from the Lolo Trail to the Clearwater Mountains.

On Saturday, September 21, one of Clark's men in the lead had returned to the main body with food and news that a Nez Perce village was but a few miles ahead in today's Weippe Prairie. Through perseverance, fortitude, grit—and Nez Perce food—the expedition had survived while covering around 160 miles in eleven days from Travelers Rest to Weippe Prairie.

History is unsure how Lolo Trail got its name. Discounting the many twists and turns, ups and downs of the trail, from Kamiah, Idaho, to Lolo Springs, Montana, the trail is 100 miles long, straight line distance. In the 1930s, the Civilian Conservation Corps began building US Forest Service Road 500, generously called Lolo Motorway. Even today this up-and-down motorway is a primitive, one lane, narrow, twisting dirt road suitable only for high clearance vehicles. The US government would not build a paved, dependable highway through this region until 1962.

In June 2024, the author partially traveled this rugged, many times hazardous, motorway in a 4-wheel drive truck having a robust V-8 engine. Throughout the demanding effort in low gear and grinding 4-wheel drive, the author thought of the Nez Perce struggling east through this tortuous, thick, timbered, and rocky region with hundreds of people, including women, children, and the elderly. They also were herding thousands of horses. Their herculean effort here cannot be dismissed by simple and complacent words of "they crossed the trail."

Weippe Prairie view north. The Corps of Discovery, or the Lewis and Clark Expedition, first encountered the Nez Perce on this prairie. It marks the western beginning or western end of the Lolo Trail.

Weippe Prairie view north.

Today's Bitterroot Valley south of Missoula, Montana.

History seems uncertain as how the nontreaty bands of the Nez Perce at Weippe Prairie decided to go to the Bitterroot Valley on the Lolo Trail. Chief Looking Glass apparently swayed the others on what to do. Three routes existed for them to cross east: The Mullan Trail stretching from Fort Walla Walla, Washington, to Fort Benton, Montana Territory; the Lolo Trail directly in front of the Nez Perce; or south across today's Chief Joseph Pass along the Idaho-Montana border.

Freight wagons used the surveyed and well-traveled Mullan Trail. This trail ran near Fort Shaw, Montana Territory. But the Army garrisoned soldiers at Fort Shaw. Going south meant a longer trail with extended effort for the Nez Perce. Following the Lolo Trail to the Bitterroot Valley and presumably friendly tribes, seemed sensible. The Nez Perce knew this trail and the country east. With Chief Looking Glass leading, the Nez Perce began the demanding journey across Lolo Trail. They crossed the trail between July 16 and July 27.

White settlers in the Bitterroot Valley had heard of the battles, killings, and rampages west across the Bitterroot Mountains. Trepidation and panic grew among the white settlers. Howard originally planned to move north to the Mullan Road and then south to block the Nez Perce. But he gave in to Army command pressure and went after the Nez Perce straight away.

At that time General William T. Sherman, Civil War veteran and architect of Sherman's March to the Sea in Georgia, was the commanding general of the US Army. Through command channels Howard ultimately

A reproduction of the enemic Fort Fizzle on Lolo Trail: a curious try at halting the Nez Perce. The Nez Perce simply went around this rudimentary log barrier.

answered to Sherman. Arguably Sherman, along with general and president, Ulysses S. Grant, were the first two modern generals in history: Modern in the way they looked at conducting war. Sherman believed in total war on his adversaries and looked at prosecuting war on the Plains Indians in that manner. Regardless of where the Nez Perce went, Sherman ordered Howard to follow and "act with vindictive earnest."

On Tuesday, July 30, Howard began crossing Lolo Trail with over 700 soldiers and civilians, along with horses and equipment: The same tortuous trail the Nez Perce had passed two weeks before. Even today these mountains are dark, a bit gloomy, with uncharacteristic moss hanging from densely packed trees perpetually wet from mist and lack of sunshine. This country is powerful and uncanny.

The Army wanted a presence in western Montana Territory southwest of Fort Shaw. In June 1877 Captain Charles C. Rawn from Fort Shaw, established Fort Missoula southwest of the town of Missoula. Rawn had heard of the Nez Perce approach and looked to block them as they came out of Lolo Pass. He led a force of over two hundred soldiers and civilian volunteers toward Lolo Pass. Near the pass they built crude barricades, two or three logs high, along the assumed Nez Perce approach path. However, in time most civilian volunteers trickled away, leaving Rawn with a scant force of around eighty.

Bitterroot Valley looking west toward Bitterroot Mountains and Lolo Trail.

On Thursday, July 26 and maybe the following day, Rawn, under a flag of truce, approached Looking Glass, White Bird, and Joseph. Rawn asked their intentions. The Nez Perce said they did not want war but wanted to pass peacefully through the Bitterroot Valley. That was good enough for Rawn as most of his force had already abandoned him. Rawn allowed the Nez Perce to pass. He then returned to Fort Missoula; his effort had "fizzled out." The Nez Perce then went north around the unrealistic log barriers, later deemed "Fort Fizzle," and went into the Bitterroot Valley. The Nez Perce camped a few miles south.

The scenic land and water rich Bitterroot Valley stretches about fifty miles north-south, and about ten miles east-west in western Montana. This is prime country and ancestral home of the Salish (or Flathead) Indians. A small band of Nez Perce, intermarrying with the Flathead, had been in this valley for some time. Lean Elk also called "Poker Joe" led this band. By their nonhostile actions in the Bitterroot Valley, the Nez Perce bands escaping Idaho across Lolo Trail thought they had left the Army and the fighting behind. Maybe the Army would not pursue them. This, of course, was not true.

The Nez Perce in the past had peaceful relations with the Flatheads. But Charlo, chief of the Flatheads, was leery of the approaching Nez Perce. Charlo had heard of battles in Idaho. He had kept a peaceful existence with the whites and soldiers in the Bitterroot Valley. Not wanting to endanger the Flatheads from any collateral military action against the Nez Perce, Charlo asked the Nez Perce to move along.

St. Mary's Mission in Stevensville, Montana. Founded 1841 by the Jesuits.

Remains of Fort Owen near today's Hamilton, Montana. Forts Owen and Skalkaho in the southern Bitterroot Valley were improvised "forts" to defend against the Nez Perce.

The Nez Perce bands had decided in Weippe Prairie to move to the buffalo country and the Crows far to the east, but how to get there? They faced three choices of going forward: Go north past Fort Missoula, but they then would be facing the fierce Blackfeet Tribe; go east from the Bitterroot Valley; or go south through the Big Hole region then east through the Yellowstone country.

Talks came of abandoning going east altogether but going north to the "old woman's country," or Canada. The views of Looking Glass, the consensus leader, apparently held sway. The Nez Perce would travel south through the Big Hole region then east: Perhaps partnering again with their supposed allies, the Crow.

News of approaching Nez Perce bands invoked stress and fear among the communities of white settlers in the Bitterroot Valley. Settlers called upon Fort Missoula for help. Coming from Fort Shaw and with soldiers from Fort Ellis, Colonel John Gibbon left Fort Missoula on Saturday, August 4, with 163 officers and men, and headed south. Gibbon was a career Army officer, a Civil War veteran who had also fought in the Plains Indian Wars. He was a fighting officer. Gibbon's objective was to overtake, capture, or fight the Nez Perce. His aim was to end this "Nez Perce business." Howard, still in the Bitterroot Valley, would follow Gibbon five days later.

Although Looking Glass and the Nez Perce showed no hostile intent in the Bitterroot Valley—citizens even traded with the Nez Perce—citizens there assembled and built rudimentary stockades at Stevensville and at today's Hamilton. The Nez Perce went past these without incident.

Looking Glass kept a slow pace up the Bitterroot Valley heading south. Climbing along a six-mile trail rising 3,000 feet from the valley, they crossed a steep 7,000-foot divide now called Gibbon's Pass and turned east. They had passed a trail, which turned back to the Salmon River watershed and to their homeland. This was their last chance to return to the ancestral home of the Nez Perce. But to these nontreaty Nez Perce there was no turning back.

The Nez Perce then descended into the scenic and immense Big Hole region west of today's Wisdom, Montana. The Nez Perce rested beneath a mountain along Ruby Creek, which empties into the North Fork of the Big Hole River.

John Gibbon, however, was not resting; he was coming hard. His horses lathered and heaving.

6

Big Hole

In the American West a "hole" is a wide-open basin in an otherwise mountainous or timbered area. Jackson Hole, Wyoming, is not a depression in the ground, but rather an expansive open area along the Teton Mountain Range. The Big Hole region in southwestern Montana is a broad, high-country, fine grass prairie in western Beaverhead Valley. This valley stretches about thirty miles north-south, and about twelve miles east-west. Beaverhead Mountains border the valley on the west and Pioneer Mountains border on the east.

The Nez Perce knew this country. This expansive valley was a path from Bitterroot Mountains east to the buffalo country. Descending Gibbon's Pass and leaving the timbered area west, the Nez Perce camped along Ruby Creek east of a hill or mountain 1,000 feet higher in elevation. Looking Glass was still consensus leader of the nontreaty bands.

Inexplicably at the Big Hole, unlike at Weippe Prairie or along Lolo Trail, Looking Glass did not keep a rear guard. These mounted warriors with fast ponies would keep an eye on things in front or in back. Apparently at the Big Hole, a few warriors wanted to form a rear guard, but Looking Glass disagreed. Why he did not agree is conjecture. In retrospect, Looking Glass violated two fundamental axioms of war: Never get caught by surprise, and never underestimate your adversary.

On Tuesday, August 7, the Nez Perce set camp along Ruby Creek: Cutting poles for lodges and digging for nutritious camas bulbs. They planned to spend a few days at this place resting and giving their horses good grass and water. Gibbon, however, was not resting. He was making good time about thirty miles a day with wagons in the Bitterroot Valley, but steep Gibbon's Pass slowed his march.

On August 7, Gibbon sent about sixty mounted soldiers on a night march to find the Nez Perce. Lead Army scouts discovered the Nez Perce camp. Scouts encountered no Nez Perce perimeter guards.

John Gibbon
(NARA)

The scouts returned to tell Gibbon. On the evening of Wednesday, August 8, he made his attack plan.

Late at night August 8, Gibbon's force of about 190 officers and men, along with one cannon, approached the Nez Perce camp from the west. Many soldiers stayed behind to guard the force's wagons and horses. Moving on foot, the Army attack force came around the hill south of the Nez Perce camp and formed a battle line along Ruby Creek and into the trees. They would strike at dawn.

During the night an elderly Nez Perce man tending his horses went for a ride. He rode directly into the soldiers. Soldiers opened fire and dropped him. The gunfire alerted the Nez Perce camp. The battle was on.

Nearing dawn, leaders in Gibbon's force had ordered their men to first give three rifle volleys into the Nez Perce, firing low into the lodges or tipis, then charge the camp. This they did. The soldier attack caught the sleeping Nez Perce camp totally off guard. But no battle plan survives first contact. The attack turned into a savage indiscriminate melee with great ferocity. Nothing appeared off limits to the soldiers: Women, children, the elderly, were savagely slaughtered. The fight became hand-to-hand with anything the Nez Perce could grab and use as weapons.

The prized Nez Perce horse herd was on the open plain near the base of the western hill north and across Ruby Creek from the Nez Perce camp. Gibbon wanted to get between the Nez Perce and their horse herd. If he could, the fight would end. But Chief Joseph saw the danger to the herd. He organized Nez Perce youth; together they saved the herd.

Gibbon had brought a 12-pound Napolean smoothbore cannon. It could fire ordnance at distances up to 1,600 yards. When the attack began, the artillery piece belched shot and flame toward the Nez Perce. But lumbering cannon rarely worked in fast-moving, sporadic, and scattered battle dynamics against Plains Indians of the American West. Nez Perce warriors killed the gun crew and captured the cannon.

Overleaf. The Big Hole battlefield. Gibbon's soldiers attacked from the left around the hill. Nez Perce village was along Ruby Creek in right of photograph. Nez Perce horses were pastured on the open area below the crest of the hill.

North flowing Ruby Creek in the Big Hole region. At the battle of the Big Hole, the Nez Perce camp was at the end of the foot path shown in right center of photograph.

The initial Army onslaught lasted about twenty minutes of ferocity unlike anything yet seen. The Nez Perce counterattacked. Their marksmanship deadly. Rifle fire cut down Army officers on the Nez Perce right side along Ruby Creek. Leaderless, the Army attack unraveled north of the Nez Perce camp. Nez Perce warriors, using brush cover along Ruby Creek, concentrated their fire on the remaining soldiers.

Army ammunition ran low. Soldiers withdrew south along the creek. The attack stalled with intermittent gunfire lasting throughout the day. Gibbon, himself, had taken part in the defense along the creek. A Nez Perce bullet had gone through Gibbon's leg and wounded his horse. Gibbon ordered a withdrawal. Gibbon now had to contend with about forty wounded men and dwindling food and ammunition. The only water was from Ruby Creek, covered by Nez Perce warriors.

While Gibbon and his soldiers withdrew, the surviving Nez Perce began leaving. They took only essentials. The fighting punished both Gibbon and the Nez Perce. Gibbon reported twenty-nine killed-in-action, and forty wounded. Nez Perce casualties vary depending on the source. Their losses appear to be between sixty and ninety killed-in-action. The number of Nez Perce wounded remains unknown.

However, the Nez Perce remain firm that a disproportionate number of Nez Perce killed and wounded were women and children. A benefit to the Nez Perce was their capturing thousands of rounds of needed ammunition from Gibbon's soldiers and packs.

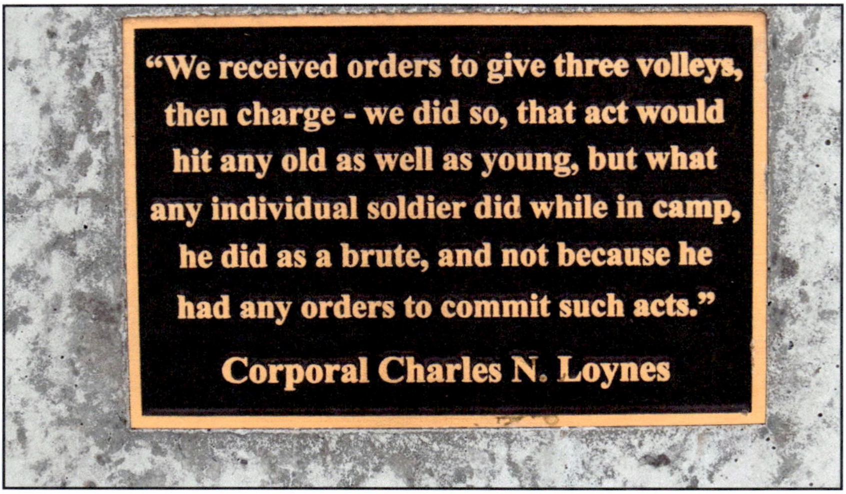

Corporal Charles N. Loynes's observaton of the wanton killing of Nez Perce women and children at the battle of the Big Hole. Plaque at the battle site.

Ruby Creek in the Big Hole region. View south. Nez Perce camp was along the foot path in lower left. Soldiers attacked from upper center around the hill at right.

View west from Ruby Creek to the hill and treeline where soldiers first formed for the attack. Gibbon placed his artillery on the hill in the treeline.

View northwest from the Nez Perce camp to the meadow on the hill where the Nez Perce horses grazed. Chief Joseph led the Nez Perce youth to save the horses.

Bulrushes along Ruby Creek. View northeast toward site of Nez Perce camp. The scrub growth concealed Nez Perce warriors as they advanced on retreating soldiers.

Ruby Creek view northest toward Nez Perce camp. The Nez Perce would bury their dead in the banks of the creek, which ran shallow at the time of the battle.

On Saturday, August 11, Howard arrived at Big Hole around late morning. Before departing, the Nez Perce had hastily buried their dead in the banks of Ruby Creek. The Army buried its dead on the battlefield. Howard began transporting his wounded to Deer Lodge.

Neither Howard nor the slightly wounded Gibbon would aggressively press the attack on the departing Nez Perce. Although the Army would state the Army had won the battle, the fight was largely a tactical draw. But as with previous fights, the Army could replenish its losses in men and materiel—the Nez Perce could not.

Nez Perce leadership was always a consensus and situational based position: Who was best to lead? Looking Glass had been the consensus headman, but after Big Hole and the Army's surprise and deadly attack, Looking Glass fell from grace. Lean Elk (Poker Joe) had joined the band near Stevensville in the Bitterroot Valley. A mixed blood, he knew a little English. He was apparently a good gambler and fierce warrior. Lean Elk knew the region and the route east. The band chose him as their consensus leader. Looking Glass stepped aside.

Lean Elk quickened the pace south. The Nez Perce had learned not to be surprised. Lean Elk sent scouts ahead and behind the Nez Perce route of march. Their march took them along the western edge of the Big Hole region and west of the former Montana territorial capital of Bannack. Along their march south, Nez Perce raiders killed several settlers and plundered ranches and farms.

Citizens at Bannack heard of the Nez Perce moving south about twenty miles west. Bannack citizens got nervous and girded their loins for protection. To reassure the citizens, Howard arrived at Bannack on Tuesday, August 14, with a group of civilian volunteers. But the Nez Perce skirted Bannack and continued south across the Continental Divide at Bannock Pass and into Idaho's Lemhi and Birch Creek valleys.

The Nez Perce arrived at the mining town of Junction, near today's Leadore, Idaho, on Monday, August 13. They looked for support from the Lemhi Band of Shoshone. But the Lemhi Band, like the Flathead of the Bitterroot Valley, told the Nez Perce to move along.

Newspapers and Army command renewed their criticism of Howard's inability to stop the Nez Perce. Howard considered his options: He could follow the Nez Perce to catch them; go east to block them at Henry's Lake or at Tacher's Pass (today's Targhee Pass) leading to newly created Yellowstone National Park. Or he could go west to block them from returning to their homeland. Howard chose to follow them.

7

Camas Meadows And Yellowstone

At the south end of Idaho's Birch Creek Valley, the Nez Perce turned northeast toward Henry's Lake and Targhee Pass. These lead to newly created Yellowstone National Park. Birch Creek Valley opens onto the Snake River Plain. Leaving Birch Creek Valley, the Nez Perce entered the dry, treeless sagebrush region of the Snake River Plain. Here Nez Perce raiders ransacked several ranches and killed a few white settlers.

Heading northeast, the Nez Perce crossed the Corrine Freight Road. This road served western Montana gold fields and began about seventy miles north of Salt Lake City, Utah, at Corrine. The Union Pacific completed the transcontinental railroad in 1869. A rail spur went north from Salt Lake City to Corrine. Teamsters transferred rail freight onto wagons, and drove about 300 miles north to Bannack, Montana Territory.

It was on this freight road, Wednesday, August 15, that Nez Perce warriors attacked mule and horse drawn freight wagons, killing seven men. Two Chinese cooks rode with the wagons and escaped with their lives that night.

Mid-August also brought heat, which dried seasonal creeks. The Nez Perce struggled, their journey hard. They had been on the run for hundreds of miles since leaving Weippe Prairie weeks earlier. Crossing Camas Creek and Spring Creek near today's Kilgore, Idaho, the Nez Perce camped south of the Continental Divide at a place called Camas Meadows.

Howard was not far behind. From the north he tried paralleling the Nez Perce march but came south to follow the well-trampled Nez Perce trail. Howard arrived at the settlement of Junction on Thursday, August 16. From scouting parties Howard learned the Nez Perce were about twenty miles east of him. Howard and his force moved east.

Camas Meadows gets its name from the camas plant found in abundance in this meadow. This meadow is a broad treeless flat area,

Centennial Mountains. View north from Camas Meadows.

having basalt outcroppings and low lava-flow ridges. Abundant grass grows along south-flowing Camas Creek and Spring Creek. The two creeks merge farther south, with Camas Creek being the combined name. Timbered mountains border this meadow east, north, and west.

On Sunday, August 19, Howard and over 200 men including Montana civilian volunteers, camped along Spring Creek and between the two creeks. He was south of the Centennial Mountains. Howard concluded, rightfully, that the Nez Perce were heading toward Henry's Lake. He sent forty forty soldiers not to Henry's Lake or Targhee Pass, but to Raynold's Pass to try and block the Nez Perce if they went that way.

The long march had depleted Howard's force and horses. They needed rest. Along Spring Creek, he camped at midday. The Nez Perce were a half-day ahead. Now it was Howard's turn to violate the two axioms of war: Never be surprised, and never underestimate your adversary.

Nez Perce scouts had seen Howard's camp and reported this to Nez Perce chiefs to the east. The chiefs decided upon a night raid to capture or scatter Howard's horses.

Mounted raids were endemic to nomadic Plains Indian tribes. Plains Indians not only raided west-moving whites, but they raided each other. These raids had multiple purposes: Acquire goods through plunder; avenge wrongs; prove mettle and merit; and achieve stature among their people. To Plains Indians, especially the Nez Perce, stealing horses meant hurting their enemies without the need to kill.

Camas Meadows

Camas Meadows, view west toward Spring Creek. Lava fields shown. Soldier positions were at the end of the foot path in center of photograph.

Camas Meadows. Today's metal silhouettes of Nez Perce warriors.

Before dawn, Monday, August 20, a Nez Perce raiding party began moving through the brush toward Howard's camp. Howard had posted sentries several hundred yards from his camp, but the Nez Perce got through. Nez Perce raiders cut ropes securing what the Nez Perce thought were horses. Then something went wrong. A single shot cracked and alerted the soldiers. The soldiers, many asleep and partially undressed for comfort, began a confused and chaotic effort to get organized to chase the Nez Perce raiders. The Army's livestock stampeded.

What the Nez Perce did not know then was in the darkness the warriors, much to their chagrin, had captured or stampeded Army pack mules and not a lot of horses.

Mules hauled Howard's vital supplies. He sent fifty men to chase the Nez Perce and recover the mules. About seven miles northeast Howard's men encountered a Nez Perce rear guard. The guard opened fire. Nez Perce warriors then split the Army force. A long-range shot killed the Army bugler. Howard brought reinforcements. By about 9:30 a.m. the fight was largely over. The Nez Perce withdrew, their casualties uncertain. Howard had eight wounded and one killed-in-action.

Again, Howard did not advance in strength and pursue the Nez Perce. Howard withdrew to Spring Creek and rested. The next day, Tuesday, August 21, he began transporting his wounded to Virginia City, Montana Territory. The forty soldiers he supposedly sent to block the Nez Perce at Henry's Lake, in fact had sent them to Raynold's Pass northwest of Henry's Lake. These forty soldiers had turned back after not

Today's Henry's Lake. View west.

seeing any Nez Perce. They missed the Nez Perce by two days.

Immediately after Camas Meadows, Bannock tribal scouts led by civilian Stanton G. Fisher, arrived from Fort Hall about 125 miles southwest. Fisher would prove a most capable scout and literate man. His journals describe in detail Fisher's observations about the effort and route in pursuing the Nez Perce.

The Nez Perce began moving toward Henry's Lake: about thirty-five miles northeast of Camas Meadows. Lean Elk (Poker Joe) remained the consensus leader among the Nez Perce bands. Chief Joseph appears to have been the camp manager of sorts, looking to camp organization and wellbeing, the youth, and the horse herd.

Henry's Lake is a natural alpine lake about 6,500 feet in altitude and covers about eight square miles. With game and fish the lake had long been an area for indigenous people. The Nez Perce stayed a short time. Howard began moving from Spring Creek.

Arriving at Henry's Lake on Thursday, August 23, Howard needed resupply and rest. While his men rested, Howard rode over sixty miles to a telegraph office at Virginia City. While purchasing supplies, via telegraph he reported to his senior commanders that he and his men had been on the march for days without rest or proper food.

General William T. Sherman had toured Yellowstone National Park from August 6 to 17. Historians speculate as to how close he came of being caught by the Nez Perce: The same adversary he and his men had been pursing for weeks.

Today's Targhee Pass looking east toward Yellowstone National Park.

Madison River, Yellowstone National Park. Both Howard and the Nez Perce followed the river when entering the national park.

Yellowstone National Park. Buffalo (bison) along Nez Perce Creek.

Nez Perce Ford along the Yellowstone River, Yellowstone National Park.

The original Mammoth hotel, lodge, or way station in Yellowstone National Park was at Mammoth Hot Springs. Nez Perce warriors shot and killed Richard Dietrich as he stood in the "hotel's" doorway.

(National Park Service)

The site of the original Mammoth Hotel is in the vale in right-center of photograph near Mammoth Hot Springs.

Mary Mountain, Yellowstone National Park. Nez Perce passed here on their way east out of the national park.

Soda Butte, Yellowstone National Park. Nez Perce passed here on their way east out of the national park.

Nez Perce silhouettes near today's Cooke City, Montana.

On August 21, Sherman was in Helena, Montana. He thought that maybe the aging, one-armed Howard might not be the man for the job. Via telegram Sherman told Howard that perhaps a younger officer might be needed to lead. Howard, sensing a veiled rebuke, replied he (Howard) was indeed up to the task. Sherman kept him.

Yellowstone National Park is America's first national park. Congress created this national park in March 1872 "for the benefit and enjoyment of the people." Yellowstone has most of the world's geysers and hydrothermal features. Indigenous peoples have been in Yellowstone for millennia. One of the first white men to visit this geological region was John Coulter. Historians debate if he ever actually went into Yellowstone. People began calling Coulter's tales of strange smells, bubbling hot water, and spewing geysers as "Coulter's Hell."

The Nez Perce entered "Coulter's Hell" on or about August 23, the day Howard arrived at Henry's Lake. The government had not yet built roads or improved trails inside park boundaries. But intrepid travelers could still get into the park.

The Yellowstone National Park chapter of the Nez Perce odyssey has mysteries still debated. Why did the Nez Perce take between ten days and two weeks to get through the park? Where they lost, or as a minimum, were they seeking other less-traveled routes east to the buffalo country? Or were they traveling slowly for rest? Again, one of the many mysteries still unanswered on the Nez Perce flight to freeom

Their route into the park from the west is reasonably known.

They followed the Madison River east (upriver) to the Firehole River. Turning south they followed the Firehole (upriver) to near Lower Geyser Basin. Near the confluence of the Firehole and today's Nez Perce Creek, the Nez Perce captured a civilian prospector, John Shivley, and used him as a captured guide through the park. He would later escape and alert Howard as to the Nez Perce presence and route.

South of where the Nez Perce captured Shivley, they encountered nine civilian tourists from Radersburg, Montana: the so-called Radersburg Party. This encounter turned violent with shots fired (one civilian wounded) and others captured, but later released or escaped. Other violent encounters with tourists happened. A party of eight men from Helena, Montana, had camped in the area. A Nez Perce raiding party found them and attacked with one man killed.

Other civilians scattered and headed north to Mammoth Hot Springs. There civilians found another Nez Perce raiding party had killed another man, Richard Deitrich. He was standing in the doorway of the rudimentary Mammoth Hotel when Nez Perce raiders killed him. Other raiders attacked north of Mammoth Hot Springs at a nearby ranch. All tolled Nez Perce raiders had killed two civilians and wounded three.

Traveling east (upstream) on Nez Perce Creek, the main Nez Perce body had passed near Mary Mountain and then crossed the north-flowing Yellowstone River at today's Nez Perce Ford. Here their trail becomes conjectured. They did not follow the Yellowstone River north and out of the park toward today's Livingston, Montana, but moved east. Following the Lamar River meant passing near the mining camp at today's Cooke City, Montana. Going east meant crossing today's Mirror Plateau, then into Hoodoo Basin, then across the Absaroka Mountains, and finally down to Clark's Fork River. Even today their exact route east is not known for certain.

Howard entered the park along the Madison River. He followed the Yellowstone River north to its confluence with the Lamar River. On Wednesday, September 5, he arrived at Baronet's Bridge over the Yellowstone River near today's Tower Junction. However, Nez Perce raiders had partially burned the bridge. Howard made repairs and continued up the Lamar River to Cooke City.

What the Nez Perce did not know was that the Army was laying a trap for them. What followed was perhaps one of the most skillful tactical maneuvers and deception in US military history. A Nez Perce maneuver that remains today astonishing for its skill, daring, and courage.

Quite simply, how did they do it? A mystery, which lingers still.

8

They Vanished

When the nontreaty Nez Perce entered Yellowstone National Park four major Army departments subordinate to two higher Army divisions found themselves involved in pursuing the nontreaty bands. Several Army generals then had something to say about pursuing the Nez Perce. Who in the Army oversaw all this was a fair question.

In the far west, Departments of the California and of the Columbia were subordinate to the Division of the Pacific. Farther east the Departments of the Platte and of the Dakota were subordinate to the Division of the Missouri. Within all these departments were district commands. All these commanders were subordinate to the Commanding General of the Army, William T. Sherman.

William Tecumseh Sherman (NARA)

When Brigadier General Howard, commander of the Department of Columbia, entered Montana from Idaho he was in the Department of Dakota. When he entered Yellowstone National Park he was in the Department of the Platte. Returning to Montana, he was again in the Department of Dakota. Technically he was out of his jurisdiction in most of his pursuit of the nontreaty Nez Perce. Apparently Sherman could care less. Sherman wanted the Nez Perce stopped, and it mattered little to Sherman as to who did the stopping.

Sherman and others had doubts about Howard's ability to stop the Nez Perce. But Howard was chasing the Nez Perce in remote areas. No other troops were available, and the Army had few options. Sherman ordered Howard to disregard all department or division boundaries and continue the pursuit.

Sherman, who had toured the Yellowstone region in August, began mustering other Army units to attack the Nez Perce. The Nez Perce were in Yellowstone National Park. The Nez Perce had three ways to leave the park without turning back to their home country: Follow the Yellowstone River north through today's Paradise Valley and Livingston, Montana; follow the miners and buffalo hunters trail along the Lamar River in Yellowstone National Park then out through Cooke City, Montana Territory, or south to the North Fork of the Shoshone River near today's Cody, Wyoming, and east to Crow country or south to the Wind River country of today's Wyoming.

Sherman, using telegraph, ordered soldiers to block each eastern route out of the national park. The fabled US Seventh Cavalry, defeated at the Little Bighorn the year before, was ordered west to central Montana and then south. Elements of the US Second Cavalry were based at Fort Ellis near today's Bozeman, Montana. They were ordered to the southern end of the Yellowstone River. The US Fifth Cavalry was ordered north along the Wind River mountains from Camp Brown, near today's Fort Washakie, Wyoming. But the voice of military wisdom and caution always whispers, "Never underestimate your adversary."

The word "Absaroka" is a Hidatsa (Sioux) Indian word for the Crow people. The word loosely translated means "children of the long beak bird." In 1877 the federal government established the Crow Agency headquarters south of today's Absarokee, Montana, northeast of Cooke City, Montana. The mountains on the eastern edge of the Yellowstone region are called the Absaroka Mountains.

Samuel D. Sturgis
(NARA)

The high Absaroka Mountains form an abrupt wall on Yellowstone National Park's eastern boundary. Many peaks in these mountains exceed 10,000-feet in elevation. These mountains form the headwaters of many streams and rivers. Even today, these water courses form the only routes through the Absarokas. These mountains and canyons are scenic but severe. A person must visit these mountains to appreciate the challenges facing the Nez Perce in getting through them.

In early September 1877 after transiting

East wall of the Absaroka Mountains north of today's Cody, Wyoming. View west from Chapman Bench along highway 120.

View west to Sunlight Basin from Dead Indian Pass in the Absaroka Mountains northwest of today's Cody, Wyoming. East exit from Yellowstone National Park. The Nez Perce came through Sunlight Basin and up Dead Indian Pass.

Nez Perce silhouettes on Dead Indian Pass, Absaroka Mountains.

Heart Mountain north of today's Cody, Wyoming. Sturgis based his forces near here.

Mouth of Clark's Fork River canyon (center in photograph) north of today's Cody, Wyoming. Sturgis positioned soldiers near here to block the Nez Perce.

Mouth of Clark's Fork River canyon and north-flowing Clark's Fork River. View south. Craggy canyon walls rise over 3,500 feet from the river.

Mouth of Clark's Fork River canyon and north-flowing Clark's Fork River. View east toward Heart Mountain, north of today's Cody, Wyoming.

Yellowstone National Park, the Nez Perce faced the formidable Absaroka Mountains. Even today few routes go through the Absarokas. At 8,000 feet, present-day Dead Indian Pass (Wyoming highway 296) on the east end of Sunlight Basin is the way east. The Clark's Fork of the Yellowstone River, named for William Clark of the Lewis and Clark Expedition, gouges through the Absarokas near the pass, carving a formidable path for the Nez Perce, but in which direction and how?

Knowing the Nez Perce would come east and possibly south from Yellowstone National Park, the Army positioned soldiers immediately east of the Absarokas near Heart Mountain. Heart Mountain is ten miles north of today's Cody, Wyoming, and the North Fork of the Shoshone River. The Shoshone River, also called the "Stinking River" named for sulfur smell from the river, is about twelve miles southeast of where the Clark's Fork comes out of the Absarokas.

Colonel Samuel D. Sturgis, a fifty-five-year-old battle veteran with experience as a Plains Indian fighter, assumed command of the US Seventh Cavalry after Custer's defeat at the Little Bighorn. Sturgis's son was killed in that battle. Sturgis was a senior officer in the Army who had seen action in the Mexican War, the Civil War, and now in the plains.

Sturgis was in the District of the Yellowstone subordinate to Colonel Nelson A. Miles, based at Tongue River Cantonment along the Yellowstone River near today's Miles City, Montana.

Miles first ordered Sturgis to go west along the Musselshell River

Bald Ridge (left center) and Bald Mountain (right center) as seen from Heart Mountain. Sturgis had his men somewhere near Heart Mountain.

View south. Today's north-flowing Pat O'Day creek is shown with Heart Mountain in upper left. The creek runs parallel to the east wall of the Absaroka Mountains and Bald Ridge. If Sturgis had his men camped along the creek, bordered on both sides by high escarpments, they could not have seen the Nez Perce on Bald Ridge.

View south. North-south Dead Indian Gulch (center) is immediately west of Bald Ridge (upper left). The gulch is about 800 feet lower than Bald Ridge, easily hiding thousands of horses and hundreds of people.The gulch goes north then curves west (lower right) toward Clark's Fork River canyon and Clark's Fork River.

View west toward craggy, deep, and aggressive Clark's Fork River canyon. The east-flowing river makes a 90-degree turn north at left-center in photograph at Devil's Gate or "the narrows." Dead Indian Gulch empties into the river in lower center shown with the sharp incised "V". However, this "V" becomes a high cliff above the river, impossible for people or horses to pass through. Two courses presented themselves for the Nez Perce if they went that way from Dead Indian Gulch: Go south (left) across the meadow and then down, or go north (right) across that meadow and then down.

Previous page. View east. Clark's Fork River canyon showing drainage from Dead Indian Gulch into the river. The gulch empties into the river at a sheer wall cliff, impossible to navigate with people or horses.

Below. Clark's Fork River canyon. View north. Dead Indian Gulch drains into the river at lower right. Impossible for people or horses to navigate.Two other possible passages from Bald Ridge to the river are north or south of Dead Indian Gulch drainage.

Clark's Fork River canyon, View south toward Dead Indian Gulch drainage and Devil's Gate or "the narrows." A possible - but not proven - route for the Nez Perce would have been for them to go north from the gulch to the passages down to the river shown in upper left in photograph.

in central Montana, ostensibly to block the Nez Perce from going north. Sturgis then went south and placed elements of the Seventh Cavalry at Heart Mountain. This was reasonable as he then could cover any Nez Perce coming out of the Absarokas going north on the Clark's Fork, or going south to the Stinking (Shoshone)River. With Howard behind the Nez Perce and Sturgis in front of them, the Army had the Nez Perce in a vice. It would appear the Nez Perce had no way out.

But all of this required precise coordination with the Army units involved. Without available telegraph these units had to use couriers, mounted or on foot, who faced the same rugged terrain as the Nez Perce.

Exiting the national park the Nez Perce followed the headwaters of east-flowing Clark's Fork River. Soon the river gouges an impassible and narrow high-walled (over 800 feet) cleft in the Absarokas, impossible to follow. They moved south to today's Sunlight Basin and to a nearly 9,000-foot hill, now called Dead Indian Hill. This hill anchors the south end of a north-south ridge in the Absarokas called Bald Ridge. This ridge overlooks the vast plains to the east, including Heart Mountain.

Sturgis relied on scouts to track the Nez Perce. This took time. The Nez Perce soon killed one scout and wounded another. September 7/8 saw Sturgis in the vicinity of the mouth of the Clark's Fork River north of Heart Mountain. Crow scouts told Sturgis that passage through Clark's Fork River canyon was impossible. Given what information he had, Sturgis went south toward the Stinking River, probably up today's Pat O'Hara Creek. He did not find the Nez Perce. He then returned on September 10 to find the Nez Perce gone from Clark's Fork River canyon.

Hundreds of Nez Perce with herds of horses, slipped away unseen to follow Clark's Fork River north to the Yellowstone River and beyond. In the Absarokas soldiers began looking in vain for the cunning and resourceful Nez Perce, but found only confusing and random horse tracks beneath Dead Indian Hill.

Bald Ridge, a smooth-topped ridge with meadows and vistas, runs north and south on the eastern wall of the Absarokas immediately east of Dead Indian Hill. Vast open plains (and the Army) were east of Bald Ridge. West of the ridge a deep valley, now called Dead Indian Gulch, runs north with high timbered sides. Hundreds of people and horses could easily hide in the gulch.

The gulch deepens farther north. If the Nez Perce used the ridge to exit the Absarokas, they had to have climbed from the gulch to the ridge. If they went this way their people and horses risked being seen on the exposed ridge's crest by soldiers near Heart Mountain. In dry September dust clouds from horses would not hide you.

Dead Indian Gulch turns northwest from Bald Ridge. The gulch narrows and like a funnel empties its drainage into Clark's Fork River through a sharp cleft in the high east wall of the severe Clark's Fork River canyon. Two possible routes west off the ridge are about 500 yards north of this cleft. These routes narrow at their peak but widen as they near the river. About 700 yards south is another probable route down to the river and out of the canyon. This route is off of Forest Trail 119.

When the Nez Perce seemed to vanish, the Army sent scouts to find their trail. The reports and memoirs of officers and chief scouts are compelling. Howard wrote in his memoirs:

> … We deviated from the usual path after clearing the mountain, and went through a strange canyon, where the rocks on each side came so near that two horses abreast together could hardly pass.

Lieutenant Charles E. S. Wood, Howard's adjutant, wrote of trailing the Nez Perce from the Absarokas:

> I shall never forget the actual pass through which he [Chief Joseph] made his exit into Clark Basin near Hart (*sic*) Mountain. It was the spout of a funnel—the dry

> bed of a mountain torrent, with such precipitous walls on either side that it was like going through a gigantic rough railroad tunnel.

Stanton G. Fisher, chief scout, and an observant articulate man left the best analysis of how the Nez Perce got through Clark's Fork Canyon. He wrote in his diary, Monday, September 10, 1877:

> After leaving the summit [Dead Indian Hill] the enemy followed the trail towards the Stinkingwater [Shoshone River] about two miles and then attempted to elude pursuit by concealing their trail. To do this, the hostiles "milled", [or] drove their ponies around in every direction, when, instead of going out of the basin in the direction they had been traveling and across an open plain, they turned short of to the north, passing along the steep side of the mountain through the timber for several miles. When we reached the point where the enemy had endeavored to cache their trail, we scattered out in every direction looking for it. At first the scouts were at a loss to know which way they had gone but after spending some time in the search I was so fortunate as to stumble onto the trail...we followed through a very narrow and rocky canon (*sic canyon*) down to Clark's Fork, at a point about two miles below where it comes out of a canon (*sic canyon.*)

Newmeyer Creek drainage somewhat fits Fisher's trail description. But it is not significantly rocky, timbered, or steep, and it narrows as it reaches the bottom of the east wall of the Absarokas. The mouth of Newmeyer Creek is about three miles south of Clark's Fork River. The other option is the timbered Dead Indian Gulch. The Nez Perce could go through the gulch almost to the cleft in the east canyon wall, and then north or south a few hundred yards to either one of the two steep drainages leading to Clark's Fork River. The canyon would then further hide the Nez Perce until the river flowed from the canyon five miles north of the cleft.

Today's Newmeyer Creek drainage (center of photograph) coming off Bald Ridge flowing east. View west toward Bald Ridge from Chapman Bench.

Today's Newmeyer Creek drainage coming off Bald Ridge flowing east. View east with Heart Mountain in upper right. This drainage could hide the Nez Perce, but it does not fit the descriptions written by Howard, Wood, and Fisher.

Paint Creek drainage is at left in photograph. Newmeyer Creek drainage is at right in photograph. Photograph taken on Chapman Bench north of today's Cody, Wyoming, not far from Heart Mountain. Given hundreds of people and thousands of horses possibly raising dust in the dry season, and not being seen by soldiers if the Nez Perce came down Newmeyer Creek drainage, remains problematic.

Seasonal drainages, now called Paint Creek and Newmeyer Creek, flow east down from Bald Ridge. Newmeyer Creek's right bank (looking downstream) flows next to an escarpment. If the Nez Perce followed Newmeyer Creek off the ridge, the escarpment could conceivably hide them from Army view. Newmeyer Creek afforded the Nez Perce a reasonable but risky path out.

Paint Creek flows east from a walled basin informally called "natural corral." This creek flows around an escarpment hiding Newmeyer Creek then north to Clark's Fork River. This, too, would be a reasonable route for the Nez Perce. But like Newmeyer Creek, the Nez Perce risked being seen by soldiers patrolling east of the Absarokas.

How the Nez Perce chose which route to follow is not known. But Nez Perce accounts say a younger Nez Perce apparently knew the area and led them through it. If the Nez Perce did indeed go north in Dead Indian Gulch, and north or south of the cleft in the Clark's Fork River east canyon wall, this would have been one of the most daring and heroic ventures in the Nez Perce War. But history will never know for certain as the Nez Perce left nary a trace of their heroic struggle.

They indeed vanished.

9

Plains And Canyons

The Nez Perce, now having around 100 warriors with the remainder being women, children, and the elderly, had gone through Yellowstone National Park, Absaroka Mountains, Clark's Fork River canyon, and now were following north-flowing Clark's Fork of the Yellowstone River. All this was a superb feat of leadership and horsemanship in severe country. The Nez Perce were now in easier-to-travel country. They had no arduous mountains or canyons left to conquer on their journey north. In September rivers ran shallow in this region, not challenging to cross.

Leaving Clark's Fork River canyon on or about September 9, the Nez Perce were in Crow country, their allies, or so the Nez Perce thought. In some manner during the Nez Perce movement north, they encountered the Crow. Accounts vary as to how this happened. Some say along Clark's Fork River, Looking Glass, relying on his earlier partnership with the Crow against the Sioux, went to meet with the Crow. He looked for cooperation. But what he found was the Crow, like the Flathead and Northern Shoshone, would not help. After Custer's defeat, the Army had brought more soldiers to buffalo country. The Crow said if the Crow helped the Nez Perce, the Crow risked retaliation from the US government.

The Nez Perce were now alone in unfamiliar land: friendless, exhausted, and depleted. The council of their chiefs met about what to do. They had few options, if any. Chiefs decided to move north to the "old woman's country," or across the "medicine line" into Canada. Sitting Bull had been an enemy of the Crow. After Custer's defeat, Sitting Bull became the unmitigated target of the Army. Sitting Bull wanted to escape the Army's continuing attacks. In early 1877 he had moved with a few thousand of his band to today's southeast Saskatchewan near Wood Mountain. Wood Mountain was a Northwest Mounted Police post at the eastern terminus of the Northwest Mounted Police Trail.

Site of Wood Mountain, Canadian Northwest Mounted Police Post on the Northwest Mounted Police Trail. Wood Mountain, Saskatchewan, Canada.

Plains Indians near the unfortified Canadian American straight-line border in the west, were not sure of the geopolitical significance of crossing the forty-nineth parallel of latitude into Canada. But what they did know was that the long knives of the blue coated American soldiers under Colonel Nelson A. Miles could not cross that medicine line and attack them. Canada could be a Plains Indian sanctuary. But the Sioux and the Nez Perce would soon discover that sanctuary did not come without a price.

Nelson Appleton Miles (NARA)

On Tuesday, September 11, 1877, Sturgis and Howard met near Heart Mountain and combined their forces. Howard assumed overall command. Moving north along the Clark's Fork River they camped at or near Bridger Bend: named for the famed mountain man and guide, Jim Bridger. They were about forty to fifty miles behind the Nez Perce. Open prairie and plains were north. A mounted force could make good

Bridger Bend along the Clark's Fork of the Yellowstone River, south of today's Bridger, Montana. Howard, Sturgis, and their soldiers camped near here.

Clark's Fork of the Yellowstone River (left) meets the Yellowstone River (right) about a mile east of today's Laurel, Montana. Photograph taken in September, the time of year the Nez Perce crossed the Yellowstone River near Laurel.

Canyon Creek empties into the Yellowstone River at today's Billings, Montana. A stagecoach station was along the creek in lower center of photograph. Nez Perce warriors ransacked the station, killed two men, and commandered a stagecoach.

time now.

The Nez Perce, with Lean Elk (Poker Joe) still as the consensus head chief, found the land smooth and undulating north of Clark's Fork River canyon and along the Clark's Fork River. The Nez Perce made good time going seventy-five miles north to the Yellowstone River. On or about Wednesday, September 12, they crossed the Yellowstone River at the confluence of the Clark's Fork River about a mile east of today's Laurel, Montana. In September the Yellowstone River runs shallow enabling an easy crossing.

Going downriver on the Yellowstone, perhaps foraging for supplies, the Nez Perce began encountering settlers. At the mouth of Canyon Creek, six miles below the mouth of the Clark's Fork River, the Nez Perce came across ranches and settlers. Running for their lives, most of these settlers got away or hid in the brush. The Nez Perce, however, did kill two men.

One settlement about a mile up Canyon Creek, was a stagecoach station built that summer. When an arriving stagecoach appeared, Nez Perce warriors overpowered the driver and took the stagecoach for a joy ride. The coach's occupants ran into the brush.

Nez Perce raiding parties continued down the Yellowstone River to the riverboat landing of Coulson near today's Billings, Montana. The

Today's Canyon Creek looking northwest from the Yellowstone River to Billings, Montana. The high valley walls are shown in the distance. On September 12, 1877, the Nez Perce camped along this creek near the north escarpment.

Nez Perce exchanged gunfire with settlers there.

The Yellowstone River has carved an eight-mile-wide valley near today's Laurel and Billings, Montana. High sandstone and chalk walls, some 300-feet high, border the valley north, farther east, and west along the river. These prominent walls offer few exits to the broad expanses of prairie north to the Canadian border.

Several rivulets flow off this wall to form Canyon Creek. The creek then flows south into an east-facing box canyon and then flows about eight miles southeast to the Yellowstone River. In September the seasonal creek remains in pools of stagnant water with alkali flats. Brush and grass border the creek bed.

How the Nez Perce knew of this creek and the exit through the north valley wall is conjecture. But they began moving up the creek toward the box canyon. They were ahead of Howard's command, but barely.

Howard knew he was behind the Nez Perce. He knew the poor condition of his men and their horses. Howard concluded he could not keep up with the Nez Perce. Howard then detailed Sturgis to take the best men Sturgis had and try to catch the Nez Perce. Howard then sent messages to the Army's Yellowstone District commander, Colonel Nelson A. Miles, US Fifth Infantry Regiment based at the Tongue River Cantonment about 150 miles east along the Yellowstone River near today's Miles City, Montana.

First contact between the Nez Perce and the US Seventh Cavalry at the battle of Canyon Creek was about four miles southeast of the canyon of Canyon Creek. The contact was at the site of a contemporary gravel pit (cleared area shown). The opening to the canyon is along the escarpment shown on the horizon.

Howard needed help and he recommended that Miles head northwest to block the Nez Perce going north to Canada. Miles, an aggressive officer, had fought the Sioux and Northern Cheyenne in frigid conditions at Wolf Mountain that previous January. Miles saw his chance. On Monday, September 17, he moved out smartly.

Earlier on September 12, the Nez Perce had camped on the broad flats up Canyon Creek from the Clark's Fork River. Crow scouts saw them and reported to Sturgis the location of the Nez Perce. The next day the Nez Perce continued up Canyon Creek to the canyon. Sturgis with not quite 800 men, pursued. The Nez Perce, with about half that number, saw the advancing soldiers and moved into the canyon.

The canyon is about a half-mile wide north-to-south, and not quite two miles deep east-to-west. The first contact between the Nez Perce and Sturgis's force was on a plateau about four miles southeast of the canyon and about two miles north of today's Yellowstone National Cemetery at Laurel, Montana.

The Nez Perce withdrew to the canyon. Warriors took up firing positions on the canyon walls north and south. The remainder of the Nez Perce and horse herd began withdrawing farther into the canyon and then moving north through an outlet in the canyon wall. This outlet follows the watershed of Canyon Creek along today's Montana highway 532, Buffalo Trail Road.

A high flat-top butte is on the southeast opening of the canyon. At one time Martha Jane Canary, the American West's colorful "Calamity

Battle of Canyon Creek, view northwest toward Horse Cache Butte (left) and canyon opening (right). Both the Nez Perce and US Seventh Cavalry approached the canyon from the southeast moving northwest.

Mouth of the canyon of Canyon Creek, view southeast along Canyon Creek. Horse Cache Butte at right. Canyon Creek in middle flowing southeast. The Nez Perce and US Seventh Cavalry approached the canyon from the southeast. Montana highway 532 (Buffalo Trail Road) in middle.

Jane," had her horses and cabin atop this butte, today called Horse Cache Butte. Sturgis advanced on the canyon with soldiers maneuvering but failing fast to get on top of this steep-sided butte and adjacent ridges. Rather than cavalry charging into the canyon and seizing the escaping horse herd, Sturgis inexplicably ordered a dismount to the dismay of his men.

The battle then became an infantry fight with soldiers trying to scale steep canyon walls and Nez Perce sharpshooters keeping the soldiers at bay. Sturgis's soldiers did in time scale Horse Cache Butte, but found the Nez Perce gone.

Crow warriors arrived at the fight, but did not engage in battle. However, they did steal many Nez Perce horses and supplies, things the Nez Perce could not afford to lose.

The fight began before noon on September 13 and ended late afternoon. The Army outnumbered the Nez Perce nearly two to one. But the Nez Perce held the high ground and used it masterfully to their advantage.

Sturgis suffered three killed-in-action and either eight or eleven wounded. History notes Calamity Jane as helping to transport wounded soldiers down the Yellowstone River to Tongue River Cantonment on flat keel boats. Sturgis reported many Nez Perce killed, but Nez Perce accounts say three warriors were wounded. One Nez Perce was killed but apparently shot dead by the Crow when he tried recovering his stolen horse.

The Battle of Canyon Creek was not as ferocious as the Nez Perce faced in other past battles—with one more to come. In fact, the Nez Perce viewed Canyon Creek as a skirmish. But it depleted them of horses, ammunition, supplies, and strength. Most importantly, it also cost them a day of travel, a day they could ill afford to lose.

The Nez Perce then went out of the canyon, probably following today's Buffalo Trail Road: the only viable route north from the canyon. Using what they could find to block the trail against any advancing soldiers, the Nez Perce left the soldiers behind. Sturgis camped that night on the east end of the canyon. His men suffered from lack of water and food.

Leaving the Canyon Creek watershed, the Nez Perce found themselves on the vast rolling plains of central Montana. Canada—and safety—was over 200 miles north.

The canyon of Canyon Creek, view west. The battle took place in this canyon. Nez Perce warriors were on canyon walls north and south (right and left in photo) and held off the US Seventh Cavalry. Sturgis had his camp along the creek at the base of Horse Cache Butte at left. The Nez Perce escaped by leaving the canyon along today's Montana highway 532, (Buffalo Trail Road) shown in middle..

Battle of Canyon Creek. Steep walls of Horse Cache Butte, view southeast. Nez Perce warriors were on the canyon walls firing at soldiers below. Sturgis had his camp along Canyon Creek at the base of Horse Cache Butte.

The canyon of Canyon Creek, view south from the north canyon wall to Horse Cache Butte. Nez Perce warriors were on both walls of the canyon firing at soldiers who were along Canyon Creek at the base of the butte. After the battle, Sturgis buried his dead along Canyon Creek at the base of the butte. Years later civilians recovered the remains of two of the soldiers. What became of the third buried soldier is lost to history.

The canyon of Canyon Creek, view north from Horse Cache Butte. Soldiers were along Canyon Creek shown in lower part of photograph. Nez Perce warriors were on both canyon walls firing at the soldiers. Present-day battlefield information kiosk is at right center at the intersection of the road and highway.

View north. The Nez Perce escape from the canyon of Canyon Creek probably followed today's Montana highway 532 (Buffalo Trail Road). The escape route is at a right angle to the east-facing canyon of Canyon Creek. The creek flowing south into the canyon is along the plateaus show at right (east).

Leaving the canyon of Canyon Creek, the Nez Perce faced the vast plains going north. Canyon Creek watershed begins at lower right. Other than river crossings, these open plains are what the Nez Perce faced in their flight to freedom in Canada. Montana highway 532 at left.

Chief Joseph statue, Laurel, Montana. Commemorating the Nez Perce chief and his people who crossed the Yellowstone River near Laurel on their unrealized flight to freedom to Canada in 1877. The battle of Canyon Creek of the Nez Perce War was fought north of Laurel, September 13, 1877. Gary Temple sculpted and cast the bronze statue. Citizens dedicated the statue in ceremony, July 4, 1989.

Annually on the Fourth of July, Laurel, Montana, holds the Chief Joseph Commemorative Run. This run honors and remembers the flight of the Nez Perce near Laurel. The long distance run and walk goes uphill north beyond Laurel, past Yellowstone National Cemetery, and near the site of the battle of Canyon Creek. Youth of the Crow Nation regularly participate.

10

North Alone

As the Nez Perce crested the high plateau north of the canyon of Canyon Creek, they faced a vast treeless prairie. The land offered no shelter or hiding place. A large natural settling basin west of today's Molt, Montana, was along the Nez Perce route of march. This basin, now called the Big Lake, holds water and attracts thousands of birds. It is conjectured that the Nez Perce camped near the Big Lake after Canyon Creek.

On September 14, Sturgis and his men having bivouacked at Canyon Creek battlefield, rode north trying to catch the resolute Nez Perce. That night after riding nearly forty miles, Sturgis camped along the Musselshell River east of today's Rygate, Montana. Trappers, hunters,

The Big Lake is a natural settling basin near today's Molt, Montana. The lake is shown at low water; Beartooth Mountains in the distance. Thousands of migratory birds rest here, which today is a National Wildlife Refuge. The Big Lake would be on the northern route of the Nez Perce. Whether they stopped here is conjecture.

The confluence of today's (diverted) Musselshell River and south-flowing Careless Creek. View north. Howard and his men camped near here. They followed Careless Creek north to the vast Judith Basin.

Indians, and the Army knew the Musselshell River. It served as a regional way point. Sturgis halted along the Musselshell for two days waiting for Howard to come up from the Yellowstone River. Howard had camped near today's Billings, Montana, at the site of the Baker Battle, August 14, 1872. Howard arrived at the Musselshell on or about September 21.

On September 17, Nelson Miles had sent a message to Howard reading Miles would take whatever forces he had and move to strike the Nez Perce. Miles was coming hard northwest from his base, the rough-hewed Tongue River Cantonment on the lower Yellowstone River. Miles had a distinct advantage over Sturgis and Howard: Miles's men and horses were fresh and well-supplied. They could and did travel fast.

Howard broke camp along the Musselshell and followed Careless Creek northwest. He had to go south around the Big Snowy Mountains to his front and then turn north through Judith Gap west of today's Lewistown, Montana.

Judith Gap is a 10-mile-wide, grassy, open space between Little Belt Mountains to the west and the Big Snowy Mountains to the east. Judith Gap and the north Judith Basin were historically prime hunting grounds. The Nez Perce went through Judith Gap and headed northeast toward the Missouri River. By this time the number of Nez Perce had dwindled. History conjectures they had a little over 700 men, women, and children. Far fewer than those were warriors.

The 10-mile-wide Judith Gap in north-central Montana. View south. The gap is between the Big Snowy Mountains east (left) and Little Belt Mountains west (right).

The vast Judith Basin in north-central Montana. View south.

Reed and Bowles Trading Post being restored along Big Spring Creek near today's Lewistown, Montana. It is conjecture if the Nez Perce traded here, but likely.

Reed and Bowles Trading Post being restored along Big Spring Creek near today's Lewistown, Montana.

Bull Creek near today's Hilger, Montana. View west toward North Moccasin Mountains.

Near today's Lewistown, Montana, the Nez Perce passed a trading post built by "Major" Alonzo S. Reed and John J. Bowles along Big Spring Creek. The Reed and Bowles trading post was on the Carrol Road. This road was an overland freight road bringing supplies west from Carrol Landing on the Missouri River to Helena, Montana Territory, and beyond. Reed and Bowles traded with anyone, including providing illegal whiskey to Indians. On Friday, September 21, the Nez Perce reached the trading post and camped there for the night.

The Nez Perce were moving north across Bull Creek to the only reasonable crossing of the Missouri River at Cow Island and Cow Island Landing in the Upper Missouri Breaks. The crossing was well known and about sixty miles northeast of today's Lewistown, Montana. The Nez Perce covered this distance in thirty-six hours.

Beginning in the early 1860s small, shallow-draft paddlewheel steamboats sailed the Upper Missouri River. During high water, river boats sailed upriver hundreds of miles to Fort Benton: the terminus of river traffic. However, this could happen but three or four months annually. During low water, boats could barely reach Cow Island and Cow Island landing: a vital river crossing about 120 miles downriver from Fort Benton. Boats unloaded supplies at Cow Island Landing. Teamsters then freighted tons of supplies north along Cow Creek Trail and west to Big Sandy and Fort Benton.

Cow Island and Cow Island Landing are on the Upper Missouri River in today's Upper Missouri Breaks National Monument in north-central Montana. The Upper Missouri Breaks, named from many "breaks" or sharp folds in the land, rank among the most isolated, unspoiled, rugged, steeply eroded, and hard-to-traverse landscapes in America. The Upper Missouri River has carved a deep trench in the Breaks for almost 200 miles. Walls hundreds of feet high flank the river. This is scenic but severe and unforgiving country.

The challenge at Cow Island and Cow Island Landing is not crossing the river, the challenge is coming off or going up the steep-walled land approaches to the river. The Nez Perce were skilled and experienced in this manner of river crossing as they had done it many times before.

Since most of the stockpiled supplies at Cow Island Landing were for the Army, twelve soldiers and a few civilians camped at the landing. The landing had no permanent buildings, only tents. The soldiers and civilians had heard of the approaching Nez Perce, and for protection dug rudimentary rifle pits and other breastworks around their camp.

On Sunday, September 23, the Nez Perce descended off the high, south plateau overlooking the Missouri River and Cow Island. Cow Island sits in the middle of a north-south bend in the Missouri River. During low water Cow Island splits the Missouri River into two channels with the island in between. The Nez Perce crossed the river and the island. They camped a mile or two north of Cow Island Landing on the Cow Creek Trail. Cow Creek Trail was the outlet to the north.

Warriors approached an Army sergeant at the landing about buying supplies stockpiled at the landing. Negotiations failed and a violent skirmish began. Casualties inflicted. The Nez Perce then took what supplies they needed, burned what remained, and moved north along Cow Creek Trail.

On September 24, a force of soldiers and civilians arrived at Cow Island Landing by boat from Fort Benton. The next day they began following the Nez Perce up Cow Creek Trail. On September 25, about ten miles north of Cow Island Landing, the Nez Perce fought another skirmish with a 13-wagon wagon train. More casualties inflicted.

The Nez Perce continued north. Eighty-five miles remained to reach Canada and safety. Thinking they had distanced the Army, the Nez Perce wanted to slow their pace. Lean Elk (Poker Joe) had been the consensus Nez Perce leader beginning at the Big Hole. He had set a steady but demanding pace of travel. In some manner, Looking Glass replaced Lean Elk as the new head chief. Chief Joseph remained a solid figure in the background caring for his two young daughters and the horse herd. Looking Glass slowed the pace, a fateful decision.

The Missouri River in the Upper Missouri Breaks National Monument. North-central Montana. View west (upriver).

The Missouri River in the Upper Missouri Breaks National Monument. North-central Montana. Cow Island shown in lower right at low water. View west. The Nez Perce most likely came off the plateau and down to Cow Island through the cleft in the breaks in upper left. This is the only reasonable approach to the river there.

The Missouri River and Cow Island in the Upper Missouri Breaks National Monument. View downriver at low water. The Nez Perce most likely crossed the river by coming through the cleft in the south wall in upper center of photograph.

The Missouri River at low water. View upriver toward Cow Island Landing. Cow Island is downriver from this photograph. The Nez Perce would have crossed downriver from Cow Island and moved north along the left bank (looking downriver or the right side of this photograph) of the Missouri River. Cow Island Landing and Cow Creek are in the upper part of the photograph at the bend of the river.

View north to Cow Island Landing at low water. Seasonal Cow Creek empties into the Missouri River as shown by the dry creek bed. Cow Creek Trail used by teamsters hauling unloaded freight, begins here and goes overland to Big Sandy and Fort Benton.

View west upriver from Cow Island Landing at low water. Seasonal Cow Creek empties into the Missouri River from the north (right in photograph). The remains of the Kipp homestead are visibile at right center in photograph.

Cow Island Landing looking downriver. The Army had a small detachment of soldiers to guard Army supplies offloaded from steamboats and stockpiled here. (Montana State Library)

Cow Island Landing and soldiers's camp. (Montana State Library)

Colonel Nelson Miles and about 500 seasoned soldiers was coming from his base at the confluence of the Tongue River and Yellowstone River. Miles arrived near today's Fork Peck Reservoir, Friday, September 21. Moving west along the Missouri River, Miles commandeered river boats to ferry him and his men along the river or across the river where needed.

Learning of the Nez Perce skirmish and ransacking at Cow Island Landing, Miles crossed the Missouri River via riverboat and headed northwest to intercept the Nez Perce somewhere near Bear Paw Mountains. By Saturday, September 29, Miles was closing in.

That same day the Nez Perce, led by Looking Glass, had reached Snake Creek, about forty-two miles directly south of the Canadian border. This seasonal creek meanders through treeless open prairie and carves a twisting path northeast of Bear Paw Mountains. The high banks of the stream afforded some shelter from biting wind and fog.

How many Nez Perce remained at Snake Creek is conjecture: perhaps five hundred with less than half as warriors. They were weary, hungry, and low on ammunition. Looking Glass did not send scouts east the direction Miles was coming—again, a fateful decision.

On the morning of Sunday, September 30, the Nez Perce heard distant thunder. A Nez Perce scout on a nearby hill signaled that an enemy approached. The thunder was not a buffalo stampede, it was Miles and his charging soldiers. The Battle of Bear Paw Mountain had begun.

The actual battlefield is about five acres in size. It is a compact place with high ground east and west. Miles, coming from the southeast, charged. Chief Joseph seeing the soldiers, hollered to save the horses northwest of the camp. Nez Perce warriors hiding in the folds and the banks of the land and the creek, opened fire with deadly accuracy. This stopped the soldiers cold. Many soldiers withdrew with casualties.

Miles had two artillery pieces: a heavy cannon, and a new lightweight Hotchkiss artillery piece. The Hotchkiss could fire an explosive shell about 1,500 yards. However, both guns were of little use given the battle situation and terrain. Nez Perce gunfire was killing the officers and leaders of Miles's force. The Nez Perce began losing leaders too. Early in the fight a bullet killed Chief Joseph's brother, then Lean Elk (Poker Joe), and then Looking Glass.

By mid-afternoon, Miles ordered an all-out assault on the Nez Perce lines, but the Nez Perce held. By late afternoon Miles concluded he had captured enough Nez Perce horses and had the Nez Perce surrounded. He also held the high ground overlooking the Nez Perce camp. But his wounded and dead covered the battlefield.

Site of the battle of Bear Paw Mountain much as it was 150 years ago. View north. Snake Creek in center. Nez Perce camp was in upper center of photograph. Nelson Miles attacked from the east (right in photograph) and from the south.

Site of the battle of Bear Paw Mountain much as it was 150 years ago. View south toward Bear Paw Mountains. Snake Creek in center. Nez Perce camp was in lower center of photograph. Nelson Miles attacked from the east (left in photograph) and from the south (upper part of photograph).

Site of the battle of Bear Paw Mountain much as it was 150 years ago. View northwest. Snake Creek in upper left. Nez Perce camp was in upper right in photograph. Nez Perce horse herd was in upper center of photograph across Snake Creek.

Throughout the fight both sides wondered if Sitting Bull and his warriors would come. They were camped about 100 miles northeast along the Northwest Mounted Police Trail in today's Saskatchewan, Canada. The Nez Perce looked north for help, while the blue coats of Nelson Miles looked north with trepidation. Sitting Bull never showed.

The next day cold and snow covered the land. Miles looked to end the fighting. He outreached to Chief Joseph. Joseph, probably unknown to Miles, was the last of two Nez Perce chiefs still alive or present on the battlefield. Joseph met Miles, but this parley accomplished little.

By evening October 1, fighting continued. At mid-morning, Wednesday, October 3, Miles ordered his artillery to open fire on the Nez Perce camp. This had little effect. The next day Howard appeared. He had been far behind the Nez Perce knowing that Miles would catch them before Howard would. Howard told Miles he (Howard) would not assume command of the battle; Howard knew Miles had carried the day.

On Friday, October 5, Miles and Chief Joseph began exchanging emissaries about discussing a surrender. During one such exchange, Howard's adjutant transcribed what Chief Joseph supposedly said as Joseph spoke no English. (The translator was most likely Arthur "Ad" Chapman.) The adjutant recorded the now famous translated words of Chief Joseph when Joseph said, "I will fight no more forever." By late morning, the battle was over. Chief Joseph surrendered his lever-action Winchester rifle to Miles.

Miles's losses were twenty-one men killed, and fifty wounded. Several of the wounded would later die. Nez Perce casualties remain uncertain. Perhaps as many as twenty-five killed and forty wounded. Both sides buried their dead with little honor. The Army buried its dead in a mass grave south of the battlefield. (In 1912 the Army reinterred the Army dead in Custer National Cemetery, Crow Agency, Montana.)

The Nez Perce War ended, but the odyssey and flight of the nontreaty Nez Perce did not end at Bear Paw Mountain. They would now travel to a place where more Nez Perce would die than all the battles of the Nez Perce War combined.

Chief Joseph's rifle displayed at the Museum of the Upper Missouri in Fort Benton, Montana.

11

Canada And Exile

As Chief Joseph and Miles stood together at Snake Creek, 418 (or 448, records vary) surviving Nez Perce, starving, wounded, ragged and destitute, came from their respective hiding places and surrendered. White Bird disagreed with Joseph as to surrendering. This had nothing to do with enmity between the two men, but rather the cultural dynamic of each band deciding for themselves what was best. One chief did not speak for all. White Bird reminded Joseph of previous battles in their homeland, and that white men could not be trusted. Joseph, wanting to save the surviving Nez Perce, men, women, children, and the elderly, had little choice but to trust Miles. White Bird planned an escape.

In the coming days and years, Joseph was adamant that he did not surrender, he said he just would not fight anymore. After Joseph's "surrender," the Army's line around the Nez Perce camp loosened. History does not record how many of White Bird's band escaped Snake Creek that night during bad weather. Estimates range between fifty to 250. They were clothed in rags, starving, and destitute. As they moved north, a few regional tribes offered help, but other enemy tribes sought to kill—and did kill—several Nez Perce.

Miles's concern was not with the escaping Nez Perce but with Sitting Bull and thousands of Sitting Bull's Lakota warriors. In late 1876 and early 1877 Sitting Bull escaped north into today's southern Saskatchewan after fighting Miles in today's north-east Montana. Sitting Bull loomed large in the Army's vengeful eyes.

Miles alerted the Canadian North West Mounted Police about the Nez Perce escape and northward movement. The North West Mounted Police, headquartered at Fort Walsh in the Cypress Hills of south-west Saskatchewan, numbered less than 300 scarlet-clad, white helmeted mounted troopers. Superintendent James Morrow Walsh commanded the troopers as they patrolled 120 miles of trail across the vast plains

Modern day historically accurate reproduction of Fort Walsh in the Cypress Hills southwest of today's Maple Creek, Saskatchewan. Canadian North West Mounted Police were based here. It was the site of the October 1877 Terry Commission with Sitting Bull. It is not known how many Nez Perce were, in time, here too.

James Morrow Walsh
North West Mounted Police
(Galt Museum)

of southern Saskatchewan. History remembers Walsh as a courageous and capable officer, looking to maintain law and order, even with escaping Indians from south of the border.

White Bird and his band found the Sioux camped at Pinto Horse Butte, about twelve miles northeast of today's Val Marie, Saskatchewan. This high butte with visibility to the horizon, divides north flowing Wood River and south flowing Frenchman's River. The Sioux welcomed the Nez Perce.

Walsh with less than ten troopers, rode into the thousands of Sioux and the arriving Nez Perce. He met White Bird and Sitting Bull. Sitting Bull had just returned from a meeting at Fort Walsh with Army general Alfred Terry, who led the so-called Terry Commission. Terry wanted the Sioux to give up and return south of the border to US government reservations. Sitting Bull, not believing any of this, rose from a buffalo robe placed for him on the floor, scoffed, departed, and never returned to Fort Walsh.

Officially the *North West Mounted Police Distinctive Monument*
Bronze statue at Fort Walsh National Historic Site, Fort Walsh, Saskatchewan, Canada. Sculpted by John B. Weaver, Anaconda, Montana. The bronze depicts a generic mounted police officer meeting an Indian on the plains. Many times it is wrongfully thought of as John M. Walsh meeting Sitting Bull.

Walsh told Sitting Bull and White Bird they could stay in Canada, but they could not cause trouble. This meant no cross-border raids or cross-border buffalo hunts. Sitting Bull and White Bird agreed.

A US Canada border marker and the 20-foot cleared swath of land between Montana and Saskatchewan south of Wood Mountain, Saskatchewan.

The Sioux and Nez Perce in Canada soon learned they were not recognized tribes in Canada. They would not receive Canadian government assistance. In Canada both the Sioux and the Nez Perce began suffering from want.

Little is known of the few Nez Perce who escaped to Canada. They journeyed to Fort Walsh and then to today's province of Alberta near Fort MacLeod. Given vast prairies and lack of identifiable land features in north-central Montana and southern Saskatchewan, a fair question is how did White Bird or Sitting Bull know they reached Canada?

In 1872 the US and Canada began

Chief Joseph and remnants of the nontreaty Nez Perce at Tongue River Cantonment (later Fort Keogh). Photograph by John H. Fouch , October 23, 1877.

a multiyear detailed survey to mark the world's longest straight-line, unfortified border between two countries. Early on, Canada wanted the US to place a marker every mile. The US said no to a maker every mile, but the US would place markers at identifiable points along the border. History does not record if Sitting Bull or White Bird knew of these markers along the border, or even if they saw them. Even today, you can find these border markers placed here and there along a 20-foot cleared swath of land between Canada and the US.

Chief Joseph at Tongue River Cantonment, October 23, 1877. Photograph by John H. Fouch.

The Nez Perce who escaped to Canada assimilated into other bands or dispersed. A few tried returning to their homeland in Idaho but were caught by whites and sent to exile or killed by tribal enemies. White Bird never returned home. One of his people shot him dead in 1892 near Pincher Creek, Alberta.

The fighting stopped; the Nez Perce War ended. Chief Joseph gathered the remaining Nez Perce and surrendered their weapons. Two days later and escorted by Miles, the surviving nontreaty Nez Perce led by Chief Joseph riding alongside Miles, began a 200-mile demanding overland journey to Tongue River Cantonment, later called Fort Keogh.

Miles and his immediate superiors agreed that the Nez Perce could winter at the cantonment. Miles told Chief Joseph that in time the Nez Perce could return to their homeland. But what Miles and Chief Joseph failed to grasp were dark clouds of changing national dynamics looming large about America's treatment of its conquered native peoples, including General Sherman's thoughts about prisoners of war.

America's newspapers and citizens began elevating Chief Joseph to an aura of distinction, leadership, and character in war. When in fact he was the only Nez Perce chief America concluded still remained.

Tuesday, October 23, Miles and Chief Joseph arrived at Tongue River Cantonment. Miles cared for the Nez Perce at the cantonment on the right bank (looking down river) of the Yellowstone River. Crude cottonwood log and mud huts characterized the base. Photographers captured images of the Nez Perce. The Nez Perce looked forward to returning home in the spring.

Howard's mission was over. Howard and his men sailed down the Missouri River to Omaha. There they boarded trains to their military bases. They had traveled thousands of miles from their bases pursuing the Nez Perce. One hundred twenty five soldiers had been killed-in-action with about 150 wounded during the Nez Perce War. The soldiers held fast and true, even when their leaders did not.

Before returning to his command of the Department of Columbia at Fort Vancouver, Washington, Howard continued east to brief his senior commanders. He advocated that the Army allow Chief Joseph to return to Joseph's homeland. But bad luck and unsympathetic press relegated Howard to a less stature than Miles.

The Army wanted out of the caring-for-conquered-Indians business. It was too expensive for the Army and commanding general Sherman could care less. He revoked any agreement made in returning Chief Joseph to Joseph's homeland. Sherman's view was the Nez Perce were just another tribe in a long line of tribes the federal government had to contend with. Sherman, a modern military general who crushed opposition, looked to have these Indians, these prisoners of war, transferred to the Department of Interior and to their monetary budgets.

Sherman ordered Chief Joseph and his band transferred to Fort Abraham Lincoln near today's Bismark, North Dakota. Sherman's view

Yellowstone River at low water looking upriver (south) at today's Glendive, Montana. The Nez Perce floating down the river would have passed the Army's Glendive Cantonment based at right in photograph, along the Glendive Road.

Commander's quarters at Fort (John) Buford twenty-five miles west of today's Williston, North Dakota, at the confluence of the Yellowstone and Missouri rivers. Sitting Bull surrendered at this house in July 1881 after four years in Canada.

Today's site of Fort (John) Buford at the confluence of the Yellowstone and Missouri rivers west of Williston, North Dakota. View southeast toward the confluence. Chief Joseph and the Nez Perce stayed at Fort Buford for a few days before moving to Fort Abraham Lincoln south of Bismark, North Dakota.

Fort Abraham Lincoln along the Missouri River south of Bismark, North Dakota. Few original buildings remain. Reproduction of Lieutenant Colonel George Armstrong Custer's house shown at right.

was to send these "prisoners of war" into captivity in the Indian Territory, today's Oklahoma. Miles protested the order in writing but to no avail. Miles told Chief Joseph that Miles had tried to keep his word, but higher commands said otherwise.

After staying at Tongue River Cantonment for less than a month, the Army began moving the Nez Perce downriver or overland to Fort Abraham Lincoln. Those Nez Perce who could, traveled about 150 miles overland along the Glendive Road in freezing arduous conditions to Fort Buford at the confluence of the Missouri and Yellowstone rivers. Both rivers run shallow in autumn, requiring skillful handling of any boat or barge to get around sand bars and fallen tree snags.

The Army loaded the Nez Perce wounded, sick, elderly, and children into fourteen flat-bottomed 8-feet by 32-feet barges called mackinaw boats, and floated them down the Yellowstone River about 175 river miles to Fort Buford. The white civilian men in charge of the boats treated the Nez Perce well. The overland journey took about ten days, again in freezing conditions. The boat journey took eight days.

The Nez Perce stayed at Fort Buford for a few days. The fort cared for them. Their journey to Fort Abraham Lincoln continued. Many Nez Perce floated down the Missouri River while Chief Joseph went overland through rugged, remote, and demanding country.

Friday, November 16, boats carrying many Nez Perce arrived at Bismark, North Dakota: a bustling city with fine buildings, cultured people, and railroads. Three days later Miles and Chief Joseph arrived. Citizens turned out in ceremony to greet Miles and the celebrated Nez Perce chief who many concluded had held at bay the best the Army could throw at him. Miles and Chief Joseph met with city leaders and newspaper reporters. The legend of Chief Joseph grew.

The Army, however, ordered Miles to St. Paul, Minnesota, and the Nez Perce south to Fort Leavenworth, Kansas. From Fort Leavenworth the Nez Perce went to the Quapaw Reservation in the Indian Territory south of Baxter Springs, Kansas. The Army ordered the Nez Perce to ride the railroads day and night through frigid conditions from Minnesota to Fort Leavenworth. They departed St. Paul, Friday, November 23.

Tuesday, November 27, the Nez Perce arrived at Fort Leavenworth to find the Army had made no preparations for them. The Army had built Fort Leavenworth in 1827 on the west bluffs overlooking the Missouri River. The fort served as a major military post and staging base for military operations in the plains. The Army reported 431 Nez Perce arrived at the fort. The fort placed them north of the fort on swampy bottom land near the river. This ground could kill as sure as a bullet.

12

The Hot Place And Home

The Army recorded 431 Nez Perce arriving at Fort Leavenworth in November 1877. By July 1878, 410 remained alive. Disease, malaria, and heat had taken took their toll. That July the Army ordered the Nez Perce south to the Quapaw Reservation in the Indian Territory where the Army had exiled Modoc Indians from the 1873 Modoc War.

The Atchison, Kansas, *Daily Globe*, July 7, 1878, announced that if any citizen wanted to see the Nez Perce before they departed Fort Leavenworth, citizens could pay 75 cents for a round-trip train ticket through the Trinity Church. A picnic was planned along with music.

The Army transferred the Nez Perce to the Department of Interior, Indian Bureau, on July 9. The Army washed its hands of Chief Joseph and the Nez Perce. The Nez Perce moved to a Fort Leavenworth railroad siding. Daily heat was intense. The Nez Perce had no shelter or water. Chief Joseph and his interpreter, Ad Chapman, became sick.

Thursday, July 21, the Nez Perce with Chief Joseph departed Fort Leavenworth on a special train to Baxter Springs, Kansas. They would pass through Fort Scott, Kansas, on their way south. Fort Scott *Daily Monitor*, Sunday July 21, recommended that the Fort Scott railroad agent have the train stop enabling citizens to look at the Nez Perce, especially Chief Joseph. The train stopped, surrounded by hundreds of onlookers. The Nez Perce suffered while sweltering in unaccustomed heat.

The train finally arrived at Baxter Springs around 7 p.m. that night. No one had planned for the arriving Nez Perce. At least six Nez Perce had died on the train. The Nez Perce buried their dead in shallow graves near the train station. In time the Nez Perce moved to a small 7,000-acre tract south along the sometimes-flooding Spring River near the Quapaw Reservation and agency in today's northeast Oklahoma. In time the Nez Perce called this new land, "the hot place." When the Nez Perce arrived, they had nothing, and nothing was waiting for them.

Today's Baxter Springs, Kansas, railroad siding. View north. Original train station was at lower right. Six Nez Perce had died on the railroad trip from Fort Leavenworth to Baxter Springs. The Nez Perce buried their dead near the railroad station. However, accounts read that something or someone disturbed the graves.

Spring River near the Quapaw Agency east of today's Quapaw, Oklahoma. View south. This is a tempermental river prone to flooding.

Quapaw Indian Agency

They shared this land with the nonwelcoming Modocs. Local citizens also were not pleased with the situation. The El Dorado, Kansas, *Republican*, May 15, 1878, wrote "… there is a conspiracy in government circles and railroad rings to have all the wild Indians of this continent settled into the Indian Territory."

The oak-tree-covered, rolling land of the Quapaw Agency was certainly not the Wallawa Valley, Chief Joseph's homeland, but it is where the Army exiled the Nez Perce. For the next seven years this region would be a demanding place where a proud and purposeful people would have to live and survive as best they could.

By August the Nez Perce situation near the Spring River on the Quapaw Reservation became critical. Disease, heat, suffering, and with few government food rations allocated by a corrupt agency chief, all came to a head. A government congressional investigation substantiated charges of agency corruption. The agency head, Hiram Jones, was sacked in disgrace. The government agreed that Chief Joseph could look for a more survivable place west in the Indian Territory.

In October 1878, Commissioner of Indian Affairs, Ezra A. Hayt, personally traveled to the Quapaw Reservation to see the situation. Hayt had earlier served on the Board of Indian Commissioners, where he had earned a reputation of integrity and being a fierce advocate of fairness with Indian tribes. Hayt arranged for himself, Chief Joseph, Ad Chapman, and others to go farther west into the Indian Territory to find a more suitable and larger place for the Nez Perce. Chief Joseph ensured Hayt knew that Joseph would not accept any land in the Indian Territory as a final home for the Nez Perce.

Leaving the Quapaw Reservation, they rode west about 150 miles across rolling land having brush and short growth juniper. Chief Joseph found suitable land bordered south by the Salt Fork of the Arkansas River, and east by the Chikaskia River.

In June 1879, a year after the Nez Perce arrived at the Quapaw Reservation, they moved west to land Chief Joseph deemed suitable. The new Indian agent at the Quapaw Agency arranged for wagons to transport the Nez Perce people. The agent accompanied the Nez Perce

Chikaskia River, east border of the Nez Perce land near today's Tonkawa, Oklahoma. View north, upriver.

Salt Fork of the Arkansas River. South border of the Nez Perce land. View west upriver.

Nez Perce Cemetery and red granite memorial, Tonkawa, Oklahoma. View northwest.

Nez Perce Cemetery and red granite memorial, Tonkawa, Oklahoma. View west.

Nez Perce Cemetery, Tonkawa, Oklahoma. Signage and memorial shown. View east.

to a place called Oakland Agency, at today's Tonkawa, Oklahoma. As the Nez Perce headed west to another unknown land, less than 400 remained.

The Nez Perce settled onto a 90,000-acre tract of tree-lined bottom land at the Oakland Agency. They did not have to share this former land with other tribes. No houses, structures, or land cultivation had been done. This land was unimproved, and the season was too far along for them to sew crops and hay. The Nez Perce endured on a primitive basis. The death rate remained high, especially for children. Chief Joseph's daughter born near Tolo Lake in 1877 was among the dead.

The Nez Perce buried their dead at a small cemetery north of today's Tonkawa, Oklahoma. Today the Tonkawa Tribe of Oklahoma, headquartered in Tonkawa, reverently cares for the Nez Perce cemetery. A white fence borders the carefully mowed and trimmed cemetery, less than a half-acre in size. Massive non-native pine trees border the cemetery to the west. Crop lands surround the cemetery. No headstones exist at the cemetery. Where Chief Joseph's daughter is buried is not known. A plaque and monument within the fence line explain the cemetery.

Throughout the Nez Perce exile, Chief Joseph began a diplomatic effort to have his people return home. He would talk with anyone who would listen as to the plight of his people. Crowds gathered wherever he spoke. And speak eloquently he did.

In March 1879 Chief Joseph and his interpreter traveled to Washington D.C. Joseph presented himself masterfully in full regalia and honor. They met with congressional committees, Department of Interior officials, and even with President Rutherford B. Hayes. Joseph made his case to allow his people to depart and go home, but nothing happened.

In 1881 Representative Richard G. Frost, D-Missouri, offered an amendment in the Indian Appropriation Bill to allow Chief Joseph and the Nez Perce to return home. Frost considered the exile of the Nez Perce as an "outrage" and hoped the US House of Representatives would "right the wrong." After a brief discussion on the amendment, it was deleted from the appropriations bill without comment.

Newspapers and benevolent organizations took up the Nez Perce cause. Editorials ranged from support, to calls for extermination of the exiled Nez Perce. Citizens in Idaho and that region did not want Chief Joseph back. They feared he would cause a Nez Perce uprising against white settlers. Keeping the peace was a priority for the whites.

By 1884 remnants of the nontreaty Nez Perce on the Oakland Agency survived as best anyone could given their situation. They farmed, raised cattle, and had a few horses. They traded with whites in Arkansas City, Kansas, to the north.

In spring 1884 citizens began petitioning Congress to allow the exiled Nez Perce to go home. These were not a few altruistic letters, but many outpourings of sentiment and support, which Congress could not ignore. Another important and influential Nez Perce supporter was (now) General Nelson A. Miles. Miles had elevated himself in the Army's and public's eyes as a man of ability and a preeminent Indian fighter who brought in Chief Joseph.

Miles had replaced Oliver Howard as commander, Department of Columbia. This department included Chief Joseph's ancestral homeland. In 1881 Miles advocated to Congress to let the Nez Perce go home. Three more years would pass before in spring 1884 Congress agreed. The nontreaty Nez Perce now numbering 283, could go home.

At the end of May 1885, Chief Joseph and his band gathered their belongings and went north about thirty miles to Arkansas City, Kansas. There they signed over any claim to their land at the Oakland Agency and boarded the railroad for their long journey home. The *Arkansas City Democrat*, June 5, 1885, reported that the Cowley County Cattle Company, Arkansas City, Kansas, acquired the Nez Perce land.

13

Epilogue

The railroad and overland journey of the nontreaty Nez Perce ended at Wallula, Washington. Here of the 283 Nez Perce who arrived, the Indian Bureau allowed 171 to go to the Nez Perce reservation at Lapwai. The remainder, 112 including Chief Joseph, would go to Nespelem and the Colville Reservation in north-central Washington. Historians debate the distribution. Issues of prevailing white sentiment and revenge, along with Christian affiliation, enter in.

The Indian Bureau did not allow Chief Joseph to permanently return to Oregon's Wallawa Valley. He continued advocating for his return. In 1897 Chief Joseph traveled to New York City and met with influential people to express his desires but without success. In 1899 and 1900 Chief Joseph visited the Wallawa Valley and his father's grave. Joseph repeated his desire to return home. This time the government was firm—he could not.

Joseph remained at Colville Reservation. He had a house there but he preferred a tipi. On September 21, 1904, while in his tipi, Joseph slumped over dead. His heart gave out.

Even in death, Chief Joseph was not allowed to go home. He was buried in a church cemetery at Nespelem. In June 1905 citizens placed a white marble monument over his grave in ceremony. An American flag flew over his grave.

Old Chief Joseph's grave along Wallawa Lake, near today's Joseph, Oregon.

"Unveiling of Monument Over Chief Joseph's Grave at Nespelem [Washington], June 20, 1905. Photo by Lee Morehouse.

(Washington State Librairies, Manuscripts, Archives and Special Collections)

Selected Sources

Books and articles

Ambrose, Stephen E. *Undaunted Courage*. New York: Simon and Schuster, 1996.

Beal, Merrill D. *I Will Fight No More Forever*. Seattle: Pullman, U of WA Press, 1963.

Brown, Mark H. *The Flight of the Nez Perce*. New York: Putnam's & Sons, 1967.

— *The Plainsmen of the Yellowstone*. Lincoln: U of NE Press, 1961.

DeVoto, Bernard ed. *The Journals of Lewis and Clark*. New York: Houghton, rev. ed. 1981

Fifer, Barbara. *Montana Battlefields*. Helena: Far Country Press, 2005.

Fisher, Stanton G., Brian V. Hunt, ed. *Indian War Diary, A Fight with Chief Joseph*. Excerpted from contributions to the Montana Historical Society, 1896.

Greene, Jerome A. *Beyond Bear Paw's*. Norman: U of OK Press, 2010.

— *Nez Perce Summer*. Helena: Montana Historical Society, 2000.

— *Yellowstone Command*. Lincoln: U of NE Press, 1991.

Josephy, Alvin M. *The Nez Perce Indians*. New York: Houghton Mifflin, 1965.

Lavender, David. *Let Me Be Free*. Norman: U of OK Press, 1992.

McWhorter, L.V., Ruth Bordin ed. *Hear Me My Chiefs*. Caldwell (ID), Caxton, 1992.

— *Yellow Wolf, His Own Story*. Caldwell (ID), Caxton, rev. ed. 2020.

Nerburn, Kent. *Chief Joseph & The Flight of the Nez Perce*. New York: Harpers, 2005.

O'Neal, Larry. *Nez Perce Exile*. Wallawa (Oregon): Bear Creek Press, 2006.

Rockwell, Ronald, *The U.S. Army in Frontier Montana*. Helena: Sweetgrass Books, 2009.

Sharfstein, Daniel J. *Thunder in the Mountains*. New York: Norton & Co., 2017.

West, Elliott. *The Last Indian War*. New York: Oxford Press, 2009.

Wilfong, Cheryl. *Following the Nez Perce Trail*. Oregon State U Press: 1990.

US Government

Department of State. *Reports Upon the Survey of the Boundary*. Wash. D.C. Gvnt Printing Office, 1878.

Newspapers.

US Library of Congress, *Chronicling America*. National Endowment for the Humanities

Internet.

Halfmoon, W. Otis (Nez Perce elder). *Lewis and Clark Through Nez Perce Eyes*. 2024 Internet interview.

Photographs and graphics.

Except where noted, all photographs and graphics by the author. For aerial photographs the author holds a current FAA remote pilot certificate and flies an FAA-registered small unmanned aircraft system in accordance with 14 CFR Part 107. At no time did the author launch from, operate from, or land upon National Park Service property. US Forest Service, Region 1, Missoula, MT, granted permission for the author to fly in Forest Service land in non-wilderness areas in Idaho and western Montana.

Index

N

O

P

Q

R

S

T

U

V

W

Y

The Author

Edward E. "Ed" Saunders was born on the Oregon Trail in western Nebraska and grew to manhood in western Wyoming. He served a career in the US (regular) Army in peacetime and war. He retired with rank of lieutenant colonel. He and his wife live in Billings, Montana.

Saunders served on the Yellowstone Historic Preservation Board, and board of directors, Western Heritage Center, Billings, Montana. He also served on the Advisory Committee for Cemeteries and Memorials, US Department of Veterans Affairs. He was instrumental in creating Yellowstone National Cemetery, Laurel, Montana. Saunders served on the national research committee to award the Congressional Gold Medal to America's women Army Signal Corps telephone operators in France in WWI, the famed "Hello Girls."

Saunders is an avid outdoorsman, photographer, and writer. He holds a current remote pilot certificate with the Federal Aviation Administration (FAA) and flies an FAA registered small unmanned aircraft system (sUAS) in accordance with 14 CFR Part 107 and applicable state and federal regulations.

"Ed, you are a white man with a red man's heart."

J.S.
Veteran, US Army
Vietnam War
Member, Choctaw Nation

Saunders on the Nez Perce Trail.

Made in the USA
Las Vegas, NV
03 November 2026

57424663R00086